Face to Face

Face to Face

*The Small-Group Experience
and Interpersonal Growth*

GERARD EGAN

Loyola University of Chicago

**BROOKS/COLE PUBLISHING COMPANY
MONTEREY, CALIFORNIA**

A Division of Wadsworth Publishing Company, Inc.

This book was edited by Konrad Kerst and designed by Jane Mitchell. It was typeset by Holmes Composition Service, San Jose, California, and printed and bound by Malloy Lithographing, Inc., Ann Arbor, Michigan.

ISBN: 0–8185–0075–1
L. C. Catalog Card No: 72–90673
Printed in the United States of America
6 7 8 9 10—77 76

Preface

More and more basic psychology courses use the small-group experience as a way of educating students to a deeper awareness of their present interpersonal style and to the whole range of interpersonal realities and possibilities. This book is a short, relatively nontechnical description and illustration of the theory underlying the small group as a means of systematic human-relations training. As such, it is meant to help the student participate more effectively in the group. In a wider sense, it is meant to serve anyone who is interested in participating in an encounter group or similar experience.

You are reading this preface because you are either involved in or interested in a small-group experience. Such groups have many names—encounter group, laboratory in interpersonal relations, sensitivity training, T-group, growth group, systematic human-relations training, and so on. But what's in a name? The group experience is only as good as what it accomplishes. Whatever the name, the members of the group sit face-to-face and speak to one another. If they can forge themselves into a supportive and understanding community—a community in which the members are basically "for" one another and come to see the world through one another's eyes—then they can train themselves to engage in the kinds of behavior that make for interpersonal growth: exploration of one's interpersonal style, a freer and more responsible expression of human emotion, a willingness to challenge others caringly and to be challenged.

Sitting face-to-face in such a small group is usually seductive,

v

intriguing, and demanding. But like most "helping" experiences it can be for better or worse, it can facilitate interpersonal growth or retard it. This book constitutes an attempt to answer the question: What kinds of behavior *facilitate* growth in the face-to-face group?

Since the book is meant to be easily understood, almost all technical references have been omitted. But this in no way implies any kind of bias against research. On the contrary, what is said here is based on the best research currently available. This book is both a partial revision and a major abridgment of my *Encounter: Group Processes for Interpersonal Growth* (Brooks/Cole, 1970). Those interested in further theory, the empirical data on which this present book is based, and complete bibliographical references should consult *Encounter*. Also, those interested in further reading in the topics introduced here may want to read *Encounter Groups: Basic Readings* (Brooks/Cole, 1971).

I would like to thank Erle Kirk of Foothill College, Donald G. Leonard of Kansas State University, Wayne Matheson of Royal Alexandra Hospital, Alberta, and Arlene Stark of the College of Marin for their reviews of the original manuscript of this book. Their comments were most helpful. Also, I would like to express my continuing gratitude to all those who have engaged in small-group experiences under my direction. They have been the source of my deepest learnings.

Gerard Egan

Contents

Face to Face

Chapter 1

Unused and Abused Human Potential

Early in the history of modern psychology, William James remarked that few men bring to bear more than about ten percent of their human potential on the problems and challenges of human living. Others since James have said the same thing, and few have challenged the accuracy of those statements. "Unused human potential" has even become the war cry of humanistically oriented behavioral scientists and philosophers.

Maslow (1968) has remarked that "what we call 'normal' in psychology is really a psychopathology of the average, so undramatic and so widely spread that we don't even notice it ordinarily" (p. 16). A great deal of human energy and money has been poured into the task of moving people from a state of "mental illness" to "mental health," but not nearly enough energy has been spent on the task of moving the mentally healthy in the direction of self-actualization. "Mental health" and "emotional adjustment" are somewhat bland terms. In a way they are like air-conditioning. Air-conditioning does not cause pleasure (except by initial contrast when it is turned on or as a status symbol). It gives relief. It renders a person's environment neutral instead of hostile and thus gives him the opportunity to make better use of his human potential—*if* he wants to. This is a big "if," however, for the unmotivated human being, despite all the investigations of the behavioral sciences, remains more or less a mystery. If a person simply does not want to relate to others more effectively, then

1

little can be done to help him. Carkhuff (1969) sees our world populated with people who are "functioning yet not fulfilling, living yet not alive, dying yet not dead." His research shows that even teachers and counselors, who are paid professionally to help others, on the average fail to do so because they lack human-relations skills. In their dealings with others they may be phony rather than genuine, vague and general rather than concrete; they may fail in basic warmth and respect for others; they may be unable to see the world through the eyes of those they are trying to teach or help; they may not know how to share themselves, to deal openly with their relationships to others, to confront lovingly and responsibly, and to suggest concrete, workable action programs that help the learner learn and the emotionally distressed find less self-destructive patterns of behavior.

Traditionally, the task of devising ways to develop a man's potentialities has been the province of education. Despite the importance of education, however, a great deal of evidence suggests that *formal* education has failed to stimulate people to grow. For example, creativity among students, far from being encouraged, is often discouraged or repressed, for the divergently thinking student tends to be a thorn in the teacher's side. Too often education, at least on primary and secondary levels in this country, is an instrument primarily of conformity rather than liberation.

The problem of the "psychopathology of the average" in interpersonal relations must be attacked at its roots. Systematic human-relations training has to accompany intellectual education in the school system. Can we look forward to the day when the curriculum of our public and private institutions, from the earliest grades on, will make the language of honest, warm dialogue a required part of general education? Carkhuff's studies (1969, 1971) provide ample proof that skills in relating well to others do not just "happen." People have to *learn* how to interact with one another. But little is done in schools to teach this kind of interaction. Children spend an enormous amount of their school time doing things next to, instead of with, others. Our society teems with this kind of "parallel" learning just as it does with "parallel" living. Therefore, it is essential, from the earliest years of education, to find ways of putting people into more effective human contact with one another. Human-relations learning is perhaps the most important kind of learning, but it is the most neglected. Schools might presume that such learning occurs naturally outside the classroom. It does not—at least not very effectively. Certainly effective interpersonal relating should be learned outside the classroom, at home and in other natural social settings. But people spend years in schools involved in all sorts of groups. Schools, then, are also natural places for training in interpersonal skills.

D-needs, B-needs, and M-needs. Maslow (1968) sees the origin of neurosis in a person's being deprived of certain satisfactions that he calls *needs*, in the same sense that water and amino acids are needs: their absence produces illness or maladjustment. Some of these basic needs are for belongingness and identification, for close love relationships, and for respect and prestige. These are D-needs (D for deficiency), for such needs, if unfulfilled, stand in the way of further human growth. Another set of needs are called B-needs (B for being); these are the needs of a person whose D-needs have been adequately satisfied but who still feels a drive for further self-actualization. For instance, a person whose basic needs have been met will probably have a need for B-love rather than D-love. D-love is selfish love; it is characterized by possessiveness, anxiety, and hostility. B-love, on the other hand, is love for the being of the other person. It is unselfish, nonpossessive, capable of being constantly deepened, and characterized by a minimum degree of anxiety and hostility.

I believe, however, that another category—M-functions (M for maintenance)—might be added to Maslow's scheme. While many people might not be grappling with marked D-problems, they still have not moved on to any significant pursuit of B-values in key areas of life such as the area of interpersonal relationships. Their relationships with others are not noticeably destructive nor are they growthful and engaging. They are bland. Their home lives are neutral—neither hot beds of neurotic interaction nor centers of interpersonal stimulation. They profess certain religious values—values that should draw them closer to their fellow men—but in practice these values are ritualistic and restraining, holding them back from doing wrong rather than impelling them to involve themselves as effective helpers in their communities. So many of us exhaust our energies in M-functions, in "keeping life going," that we have little energy left over for B-functions.

Encounter Groups in Terms of D- and B-needs and M-functions. Is there such a thing as an encounter group where the participants come together to pursue only B-needs? Bugental and Tannenbaum (1965) tried to do just this; they arranged a laboratory called "Advanced Sensitivity Training." Members were chosen because of "functional excellence" in vocation, marriage, and friendship, because they manifested an observing and curious ego, and because they were highly motivated to participate in group experiences designed to foster human growth. Also, they gave evidence of possessing adequate tolerance for psychic stress arising from ambiguity, intrapsychic conflict, interpersonal conflict, uncertainty, and risk. But the group experience didn't work. The idea looked great on paper, but the practice was something else. For all the self-actualization of the people who constituted this group, they were still beset by a clear range of emotional interferences with their functioning. They preferred to recognize and deal with the negative and the pathologic within themselves and were unsure and self-conscious about their positive and creative resources. The group leaders found themselves time and again more active in pointing out distortions in human functioning than encouraging growth, risk-taking, and creativity. It is easier to point out to a person that he has large reservoirs of anger bottled up inside than to teach him to be more empathic in his dealings with others. I think there are at least two lessons to be learned from this experiment: (1) encounter groups must deal with the whole man, and every man, even those most engaged in B-functions, must grapple to some extent with D-needs and M-functions all his life, and (2) human-relations training is most effective when it is programmed—that is, when the goals of the group experience are clearly spelled out in operational, behavioral terms and when the means for achieving these goals are similarly concrete and specific. The following chapters, then, are meant to provide the basis for a systematic program of human-relations training.

If an encounter group were composed principally of participants with severe, unresolved D-needs, then it would be much more similar to group therapy than to a laboratory in interpersonal relations. Encounter groups, then, are usually made up of participants with a mixture of D-problems, various degrees of M-function overinvolvement, and B-aspirations and skills. D-needs are not the overriding concerns of the group, but nearly all participants express some D-concerns. A more important focus in encounter groups is M-overinvolvement. It is the person who is overcommitted to M-operations in his personal and interpersonal living who is the principal victim of the "psychopathology of the average." One of the functions of the laboratory is to "unfreeze" or loosen up the M-person, especially in the area of interpersonal living.

An article in a nationally circulated newspaper reported that some businesses were dropping sensitivity-training programs for their managerial personnel. It seems that many managers went off to the labs and were "unfrozen." But when they returned to the job, they looked around and said: "This isn't for me." What businesses need are total organizational development *programs* and not merely "unfrozen" managers. Herzberg (1968), for instance, has discovered that with respect to human motivation in organizations there are two sets of factors operating. Good pay, good hours, good company policy, good relationships with supervisors, and like factors are, in his terms, *hygienic* factors. They make the organization a "clean" place in which to work, but they do not motivate the employee to put forth his best effort. Achievement, recognition for achievement, the work itself, responsibility, and growth or advancement are the principal *motivators*. Hygienic factors cut down on job *dissatisfaction*, but motivators contribute to job *satisfaction*. Organizations can benefit from programs that include both human-relations training and job-enrichment (increasing motivators) programs, but the indiscriminate use of sensitivity-training experiences to produce "unfrozen" managers is a disservice to the organization and perhaps to the managers. Both managers and organizations may well need unfreezing, but these needs can best be handled through comprehensive, goal-directed programs.

Chapter 2

Laboratory Learning
and the Encounter Group

The encounter group, as I conceive it, is a species within a much wider genus called laboratory learning or laboratory training. To put it quite roughly, *laboratory learning* is a form of experience-based education in which the participants learn principally through some sort of activity or practice rather than from lectures or books. The participants come together in small, face-to-face groups in order to interact with and receive feedback from one another in ways that have been proved to develop a variety of human relations skills. Each member, by reflecting on his own behavior and by means of the feedback he receives from the other members, has the opportunity to get a feeling for ("diagnose"), experiment with, and improve his interactional or human-relations style. For instance, if the group is composed of managers, they engage in a set of structured experiences or exercises through which they express a wide variety of managerial behaviors in order to diagnose and improve managerial behavior. One such structured experience is called "hidden agendas." Some of the participants are given certain roles to play in a mock board meeting. The board has the task of making a decision about a possible merger—this is the *manifest* agenda of the board. However, the roles given to various individuals also include *hidden* agendas that obstruct the work of the group (for example, the role that one person receives says that he is always to fight against whatever suggestions a certain other member offers). Not all of the laboratory participants play roles in the mock board meeting; some act as observers. After the board "meets" for a while, time is called and the observers give feedback to individual members with respect to how well they played their roles and what they saw happening in the meeting. The hidden agendas of the various role-playing members are then revealed and the group discusses how the hidden agendas interfered with the manifest agenda of the group. The participants, therefore, learn by doing and by receiving feedback on their behavior.

The situation in which they find themselves is called a laboratory for a number of reasons. The participants "research" their own behavior and they "experiment" with "new" (hopefully more growthful or goal-directed) behavior. It is also a laboratory because the setting in which they work is artificial rather than natural. Evidently such laboratory learning is most appropriate for situations in which human relations are critical—for instance, basic interpersonal relations, education, management, the helping professions, community relations, organizational development, black-white relations, and the consultation process.

The encounter group is a *specific* form of laboratory learning. The focus of this laboratory is interpersonal relations as such (rather than, for instance, human-relations skills in the context of managerial behavior or human-relations skills insofar as they relate to a community-development situation). A small group of people come together to assess their interpersonal strengths and deficits (diagnosis) and to experiment with effective forms of relating that have not been part of their day-to-day interactional style. The encounter group, then, is one species of the genus *laboratory learning*. In the present chapter we will take a brief look at the elements of this genus. The rest of the book will deal with the species *the encounter group*, a laboratory in interpersonal relations.

The Genus: Laboratory Learning

Someone once said that it is better to feel compunction than to know how to define it. This sentiment, paraphrased and applied to laboratory learning, might read: It is easier to experience laboratory learning than to describe it. With this caution, let's look at the elements common to most, if not all, laboratory experiences.

Learning through Actual Experience in the Small Group.   In most laboratories the most important learning takes place through the interactions in the small, face-to-face conversation group itself. Learning may also take place through independent reading, lectures, and exercises that focus on various aspects of the group experience, but such learning is secondary; it is important to the degree that it enhances the quality of interactions in the small group itself. Since the most important input is the behavior of the participants themselves, all laboratory experiences have a strong "here-and-now" flavor. The participants are learning how to learn from the behavior they themselves produce during group sessions. For instance, a group of counselor trainees learn cognitively that the more genuine they are and the more concrete they are in dealing with the counselee's experience and behavior the more likely are they to be of help. They then interact with one another in a group experience and are rated by observers on the concreteness and genuineness of their responses to one another. They also give feedback to one another on how they experienced one another in the dimensions of concreteness and genuineness.

For most of the participants in a laboratory in interpersonal relations, this kind of learning is usually quite different from the kind they are used to. The whole purpose of the laboratory is to translate good theory into good action, or to learn good theory through action. Since the participants are often unprepared for this kind of action-learning, they take flight from the process in various ways. One of the principal modes of flight is to engage in discussions about abstract concepts and values. But the laboratory demands interpersonal performance rather than discussion. Even if the laboratory is an academic course, books are secondary. If there are lectures, they are short. The general authoritarian structure associated with learning is set aside. In fact, since many of the cues that traditionally enable the "learner" to identify a learning situation are absent (no long lectures, no note-taking, no outside reading), some participants are disturbed. But if the goals of such experiential learning are spelled out carefully in behavioral terms ("the goal of this exercise is to help you differentiate between a concrete statement about your relationship to a member of the group and a general or vague statement"), then this learning can be extremely rewarding.

Cultural Permission. Laboratories are sometimes referred to as "cultural islands," both because they are cut off or insulated from the highly routinized culture of the back-home setting and because they develop their own culture in miniature. If the laboratory is to succeed, it must have a cultural freedom or permission that is lacking outside the group. This cultural permission is one of the keys to the success of laboratory learning: it allows the laboratory situation to be different from day-to-day living so that it might make a difference in such living.

What are some of the cultural permissions afforded by the laboratory? It allows comparative strangers to talk with one another at often deep levels of intimacy: the cultural prerequisites for friendship and intimacy are laid aside in so far as this is possible. The participants in the encounter group deal with one another intimately, not because they are longtime acquaintances but merely because they are fellow human beings. Confrontation is another important area of cultural permission. We seldom tell one another what impact we have on one another. In our culture, it seems much more permissible to tell a *third party* impressions about another that we would not dare tell the other, even if he is a good friend. We call this serious business "gossip." The encounter group allows the participants to disclose themselves, express their feelings, evaluate the behavior of others objectively and directly, lay aside those forms of politeness that are really nothing more than constructions that make relating "safe," and do anything else that is both interpersonally responsible and interpersonally growthful. There is some complaint that some laboratories go too far, that they permit or even encourage too much, so that the cultural and even the ethical sensibilities of the participants are offended. In certain cases this is true, but it need not be the case. This is no reason to condemn laboratory learning or encounter groups as such. Responsible laboratories encourage cultural permissiveness without subscribing to ethical license.

A Climate of Experimentation. Theoretically, if the group experience is to have an impact on behavior outside the laboratory, if it is to make a difference in day-to-day living, then it must be *different* from day-to-day experience. The interactions in some way must dramatize the overlooked dimensions of the kinds of behavior that are the focus of the laboratory, such as managerial styles, group decision making, and interpersonal relations. Therefore, participants are encouraged to experiment with their behavior during the laboratory—that is, attempt kinds of behavior that previously have not characterized their style. The climate of cultural freedom mentioned above helps build an atmosphere in which such experimentation is natural. In some laboratories, exercises are introduced to shake the participants out of routine ways of acting. Admittedly, this is anxiety arousing, but anxiety itself, if kept within limits, becomes a stimulus to new forms of behavior.

Group Size.   The group has to be small enough to allow each participant the opportunity to contribute to the interaction, but it must be large enough to allow the participants to space their contributions. If the group is too small, each member is constantly "on call"; if the group is too large, it is too easy for the individual to hide in the crowd. Also, the group must be large enough so that there is some heterogeneity of contribution, diversity of opinion, coalition formation, and other variables that are considered essential to the learning goals of the group. Another way to view group size is to say that the group should be large enough so that the absence of one or two members does not debilitate the group and yet small enough so that such absences are felt. In practice, laboratory groups usually range in size from eight to sixteen members, but optimal size is determined by the nature of the group and its goals. In encounter groups, I personally prefer about nine members. In such a group each member usually has ample opportunity to try to establish a relationship of some closeness with each of the other members of the group.

Feedback.   Group members, both individually and corporately, reflect on the behavior in which they are engaged—problem-solving work sessions, group decision-making conferences, managerial-planning sessions, the establishment of relationships of some closeness, or whatever kind of behavior is the object of the laboratory—and share their reflections. The group "processes" its own behavior as a group and the behavior of the individuals in the group in terms of the learning goals of the group. For instance, let us say that the facilitator in an encounter group interrupts the interaction of the group and suggests that the members process their interaction with one another. A member might say: "We're really not very spontaneous with one another. I get the idea that everyone is being very cautious. I know that I'm being cautious because I'm not sure that everyone here accepts me. In fact, Jim, I think I'm most afraid of you, because for some reason I get the feeling that you really don't like me." Feedback requires initiative on the part of the members, initiative to speak up. The group that cannot develop a climate of initiative will not learn very much.

Leadership. In laboratory experiences the leader is usually called a "trainer" or "facilitator" and acts, ideally, both as a model and as a resource person for the group rather than as an authoritarian figure. He models the kinds of behavior that help members achieve the stipulated goals of the group. He also facilitates the examination and understanding of the experiences of the group: he helps participants focus on the way the group is working (or not working), the style of each individual's participation (or nonparticipation), and the issues that are facing the group (or that the group is not facing). As a participant-observer, the facilitator attempts to reveal to the group its own dynamics and to clarify whether the individual participants are or are not achieving the goals of the group. In practice, there are a wide variety of leadership styles, each trainer differing markedly from the next with respect to such variables as frequency of intervention, directive tendencies, degree of self-involvement and self-revelation, and depth of confrontation. Chapter 5 will deal in greater detail with the leadership process that is considered most effective in encounter groups.

Communication and Emotion. In almost every laboratory one of the principal emphases is the network of problems centering around communication processes. Participants learn, often painfully, that it is impossible to deal with issues—even in a laboratory concerned with effectiveness in decision-making processes—on a purely intellectual level. Because group members have a tendency to ignore emotional and interpersonal issues, these issues can become the major barrier to task effectiveness in many groups. Let us take an example from an encounter-group lab in which one of the goals is to have each member try to establish and develop a relationship of some closeness with each of the other members of the group:

Avoidance behavior	*Contactive behavior*
Jim: I would like to get to know you better, but I don't seem to know how. *Jane:* We both seem to be having the same problem, but then everybody can't get in touch with everybody else in the same way and to the same degree.	*Jim:* Jane, you and I avoid each other in this group. At least I actively avoid you and you don't seem to contact me. I'm a bit afraid of you. *Jane:* I wondered when this would come out into the open. You deal so freely with almost everyone else here. I feel I have the plague in your eyes. *Jim:* I feel attracted to you and I keep tripping over my emotions. I don't want to be rejected by you. *Jane:* We are both sitting here locked up in ourselves like dodos, each fearing rejection from the other.

Support.   A laboratory is an opportunity for the participants responsibly to lower the defenses that tend to rigidify the personality and distort reality. But since the laboratory deals with emotional issues and tends to shake the participant loose from accustomed patterns of acting, it must also provide a climate of support to sustain the kinds of interactions it demands. This is one area in which some laboratories fail: they demand anxiety-arousing behavior from the participants and yet do not provide an adequate climate of psychological safety. The best assurance of psychological safety is the existence within the group of a strong climate of mutual respect and support. The facilitator especially should both model supportive behavior and prevent the group from clawing at any particular member.   ,

Exercises. Exercises are used in many laboratories to stimulate participation, to introduce missing elements into the group experience, and to highlight various aspects of participant behavior. For instance, in the encounter group there might be an initial data-sharing exercise. Each member is asked to think of something he likes about himself and something he dislikes about himself that might affect his participation in the group. For example:

> *John:* I find it difficult to take initiative in groups. This is something I dislike. On the other hand, I'm a warm person. You don't have to prove yourself to me before I accept you.
>
> *Mary:* I'm impatient with people and often don't give others a chance. As a result I'm a poor listener. On the other hand, I'm open and honest. You'll know where you stand with me.
>
> *Jane:* I'm extremely shy. That's one of the reasons why I'm here. Paradoxically, I seem to deal more freely with emotions than some. I'm not afraid to show my weakness by crying and I'm not afraid of showing affection.

This data pool becomes the basis for interaction. Since each participant speaks and shares himself to some degree, it is possible for any given participant to contact any other participant directly about a here-and-now issue.

> *Jane:* John, you said that you find it difficult to take initiative in groups. I'm the same way. Maybe we can make sure that neither of us sits back and just watches the interaction. You help me and I'll help you.

Exercises may be either verbal or nonverbal. However, some groups become too dependent on exercises, and their interactions come to lack spontaneity.

Suspension of Judgment. If the laboratory offers the participant a climate of interpersonal freedom that is available in few other social contexts, it also demands from him an openness to the experiences of the laboratory. Some participants come to the laboratory having prejudged either the entire experience or significant parts of it. For instance, some come saying that they refuse to participate in exercises because they are too "phony." Often they really mean that they see exercises as too anxiety-arousing. All laboratories involve a certain amount of risk, but life itself, if it is to be fully lived, entails a great deal of risk. The question is not whether the encounter group involves risk but whether the risk is worth taking. Risk engenders anxiety and anxiety engenders defensiveness. If the participant realizes before entering the laboratory experience that he may be the victim of excessive defensiveness, then he may be prepared to handle it creatively. This hardly means that he should hand himself over blindly to experiences that might depersonalize him, but it does mean that he should develop an attitude of responsible risk-taking.

Artificiality-Reality Dimensions of Laboratory Life

At first glance it would seem that a laboratory in interpersonal relations labors under a relatively high degree of artificiality. In a sense this is true; all laboratories are contrived. But the question is not whether they are contrived but whether they train participants in the skills that are the focus of the laboratory. The success of the laboratory for each individual is determined not by the process but by the outcome.

Laboratory Artificiality-Reality. Let's take a look at some of the artificialities of laboratory learning. First of all, the participants do not come together "naturally"; they are quite often strangers. And yet they are expected to achieve quickly a certain degree of intimacy in the give-and-take of group interaction. The artificiality of being with members of any particular group (*this* group) is often emphasized if the laboratory is composed of a number of groups, for after a while certain members begin to feel the "distant-fields" urge, saying to themselves "I wish I were in that other (good) group."

On the other hand, one of the assumptions underlying the laboratory experience is that it must be different from day-to-day living if it is to make a difference in the life of the participant. The artificialities of the laboratory in interpersonal relations are valuable to the degree that they highlight overlooked realities in day-to-day interpersonal living. For instance, we live in a society that is highly mobile and becoming increasingly so. In such a society the ability to establish relationships of some meaning and closeness quickly is essential if anomie and isolation are to be avoided. Although Aristotle suggests that people cannot become friends until they have eaten a "peck of salt" together, and although the laboratory does not espouse "instant intimacy" as a cultural value, it does allow the participant to discover that it *is* possible to make meaningful human contact with strangers in a relatively short time.

The participant involves himself with this group of people not *because* they are strangers but because they are human beings. By joining a group freely, the participant locks himself into *this* group and must face *this* set of interpersonal realities. Even though we live in a highly mobile society, it is also one of the realities of human living that people *are* locked into relationships with particular individuals and groups. In the laboratory, interpersonal problems cannot be solved by ignoring them, by moving to a different group of persons, or by using other modes of interpersonal flight. The pressure for involvement with this particular set of people, while artificial in one sense, highlights unproductive modes of involvement or noninvolvement with the "real" people in the participant's normal life situation. Moreover, dealing with the stranger in the laboratory can bring home to the participant, in a dramatic way, his failure to deal with the "stranger" element in those with whom he is intimate in real life. The laboratory, then, does not allow the opportunities for flight from intimacy that day-to-day living often does—or at least these tendencies to flight are openly challenged. Therefore, part of its impact arises from its being *more* rather than less real than ordinary interpersonal living.

The encounter-group laboratory forces the (willing) participant not only to deal with a certain set of people but to do so in specified ways. In a high-level laboratory experience, the participant learns how to talk about himself, how to reveal the "person inside" more responsibly, how to foster constructive emotions and handle destructive ones, how to show care and concern for others, how to challenge others with care and involvement, how to see the world through the eyes of others, how to understand others, how to engage in self-exploration that is neither exhibitionistic nor masochistic—in a word, how to be a more fully functioning human being. The member, then, finds himself engaged in interpersonal activities that may not be a part of his day-to-day interpersonal style. A participant, in his "real" life, might manage to avoid letting others, even friends, know about the "person inside"; he may court peace at any price in his contacts with others and thus avoid confrontation and the self-examination the confrontational process involves. But the pressures of the encounter group force him to face, at least to some degree, these realities of interpersonal living.

Exercise Artificiality-Reality. Exercises or structured experiences are sometimes used in laboratories to help stimulate the participants to make deeper contact with themselves and others. For instance, the "unanswered-questions" exercise might be used after the participants have been together for a number of sessions. During the time of this exercise the participants can interact only by asking questions of one another. Therefore, someone begins by asking a question of someone else. The participant who was asked the question then asks someone else a question (he may not immediately ask the person who asked him). Nobody answers any of the questions. The purpose of the exercise is twofold. First of all, the participants begin to learn that questions are very often statements that are disguised as questions. Note the following examples:

The Question	*The Statement*
How do you feel now?	You look bored to me.
Why do you keep picking on John?	You pick on John unfairly and it frustrates and angers me.

What can I do to let you know me better?	I'm afraid to let others inside to know me; you scare me especially because I feel you are a controlling person.
Why don't you like me?	I feel bad that you don't like me because I do like you.

When the participants are artificially restricted to question-asking, they soon see that they can translate most of their questions into concrete, direct statements. Second, since the participants do not have to fear an immediate answer, they usually begin asking bolder questions (that is, they make bolder statements, statements that get at interpersonal issues that have not been surfaced in the group). Exercises, then, both verbal and nonverbal, are often used to stimulate different modes of communication or to dissolve communication blocks that arise in the group. Whatever their purpose, however, they are still artificial; they are ways of acting that are not current in even relatively intimate associations of day-to-day living.

On the other hand, exercises are potential sources of a great deal of interpersonal reality. Exercises focus on molecular aspects of interpersonal relating (such as question-asking or eye contact) in order to make them more real in molar living, just as some artists exaggerate forms in their painting in order to make the observer look at a form that has never really lived in the observer's eye or to revive a feeling for this form that once may have lived but is now dormant. In a sense, exercises are the "zoom lens" of the laboratory experience. If they are used judiciously, they can make certain aspects of interpersonal relating come alive in dramatically new ways.

In summary, the artificialities of the laboratory are meaningful to the degree that they are useful in changing interpersonal attitudes and behaviors in the direction of fuller interpersonal living. Much of what takes place in the laboratory is artificial only in the sense that it is not what is usually done in interpersonal relationships, not in the sense that it is false or inauthentic. Laboratory groups, then, are something more and something less than real life. They are certainly more meaningful than the over-ritualistic and cliché relating that goes on in everyday life, but they are less meaningful than the natural, spontaneous, growthful contacts that take place between those who choose one another as friends and who are not afraid to relate to one another at the deepest levels. Perhaps the laboratory is best seen as a training session in human relations: the participant has the opportunity to examine his level of functioning in interpersonal situations, to become a resource person in those areas in which he functions at high levels, and to become a learner in those areas in which he functions at a low level.

Laboratory can be an orientation toward life. If I am laboratory-oriented, this means that I, like Plato, find that the unexamined life is not worth living. I am aware of my areas of strength in human living and my areas of deficit. I take risks in order to decrease whatever interpersonal dissonance there is in my life and to actualize—in the community of my friends—my possibilities, interpersonal and otherwise. It means that I struggle to avoid both dependence and counterdependence and opt for interdependence with others, realizing that others have resources for my own growth which they are willing to share if I am willing to share my own.

Chapter 3

The Importance of Goals

Listed below are a number of behaviors that are essential to high-level human relating. Rate yourself on these behaviors, using the following scale.

1	/	2	/	3	/	4	/	5	/	6	/	7	/	8	/	9

Very weak	Moderately weak	Adequate	Moderately strong	Very strong

Note that a rating of 5 means that in a particular category you would consider yourself a resource person (if only minimally so) in a human relationship or a group, a giver in that category rather than just a receiver.

___8___ *Empathy:* I see the world through the eyes of others; I understand others because I can get inside the skin of others; I listen well to all the cues, both verbal and nonverbal, that the other emits and I respond to these cues.

___8___ *Warmth, respect:* I express (and not just feel) in a variety of ways that I am "for" others, that I respect them; I accept others even though I do not necessarily approve what they do; I am an *actively* supportive person.

___6___ *Genuineness:* I am genuine rather than phony in my interactions; I do not hide behind roles or facades; others know where I stand; I am myself in my interactions.

19

___6___ *Concreteness:* I am not vague when I speak to others; I do not speak in generalities nor do I beat around the bush; I deal with concrete experience and behavior when I talk; I am direct and specific.

___6___ *Initiative:* In my relationships I act rather than just react; I go out to contact others without waiting to be contacted; I am spontaneous; I take initiative over a wide variety of ways of relating to others; when in a group I "own" the interactions that take place between other members and get involved in them.

___8___ *Immediacy:* I deal openly and directly with my relationships to others; I know where I stand with others and they know where they stand with me because I deal with the relationship.

___9___ *Self-disclosure:* I let others know the "person inside"; I am not exhibitionistic, but I use self-disclosure to help establish sound relationships with others; I am open without being a "secret-revealer" or a "secret-searcher," for *I* am important, not just my secrets.

___8___ *Feelings and emotions:* I am not afraid to deal directly with emotion, my own or other's; in my relationships I allow myself both to feel and give expression to what I feel; I expect others to do the same; but I do not "inflict" my emotions on others.

___4___ *Confrontation:* I challenge others responsibly and with care; I use confrontation as a way of getting involved with others; I do not use confrontation to punish.

___5___ *Self-exploration:* I examine my lifestyle and behavior and want others to help me do so; I respond to confrontation as nondefensively as possible; I am open to changing my behavior. I use confrontation as an opportunity for self-exploration.

Rate yourself and have others rate you. See whether others see you as you see yourself. After the group experience, see whether you and/or others see you as improved in these behaviors. Encounter groups are laboratories in which you have the opportunity to develop or increase the skills listed in the scale above. Developing and increasing these skills constitute the goals of the group. Some proponents of encounter groups think that the goals of the group need not be concrete, clear, and specific. This point is important enough to be dealt with separately.

Ambiguity Versus Clarity of Goals

Peter Drucker (1968) in his *The Age of Discontinuity* suggests that one reason why certain organizations are not as successful as they might be is that they lack clear-cut criteria by which they can judge success and failure in the organization. To oversimplify, businesses have a clear criterion of success and failure. If at the end of the fiscal year the books are in the black, the business has succeeded, more of less; if the books run red, then there has been failure of one degree or another. Other organizations, however—government, education, and church are good examples—do not have such clear-cut criteria. At the end of the high school year, how is the school to determine whether it has succeeded or not? What are the criteria for judging the success or the failure of a parish or other religious congregation? These questions are difficult to answer, for the goals of these organizations are often so vague that it is impossible to establish concrete criteria to evaluate these organizations. There is an analogous problem in laboratories in interpersonal relations.

Confusion in Face-to-Face Groups. Encounter groups are not the only kinds of groups that suffer from at least initial confusion. Writers have begun to point out that the same is true of psychotherapy, whether it takes place in the smallest kind of group, one-to-one therapy; or in what is commonly known as group therapy. The average therapist does not tell the client anything about the nature of psychotherapy, what he, the therapist, will do, and what he expects the client to do. The average client must buy a pig-in-a-poke and he is lucky if it turns out to be a succulent pig. All research, however, suggests that giving clients prior information about the nature of psychotherapy, the theories underlying it, and the techniques to be used will facilitate progress in psychotherapy. More and more practitioners are finding it scandalous that therapists are unwilling to impart to their clients much about the process of therapy.

Clarity of Goals in Encounter Groups. In the average encounter group the participants receive little if any knowledge about the goals of the group, the function of the facilitator, and the techniques and processes to be used to pursue whatever goals there are. For some, then, goallessness or "planned ambiguity" is part and parcel of laboratory training. According to them, the group *must* start goalless because one of the principal functions of the group is to *create* its own goals. This is fine if one of the goals of the group is to mill around and search for goals. It is not acceptable if the purpose of the group is to help the participants increase their interpersonal skills.

There is a great deal of evidence indicating that groups achieve a high degree of effectiveness or operationality only through clear goals and carefully defined means for achieving these goals. A nonoperational goal is one that is quite general in itself (for instance, "to become a group," or "to become sensitive to others") and one that is not realized by a particular sequence of group activities. If effective human-relations training is to take place through encounter groups, then the goals of the group must be clear and the means to achieve these goals must also be clear.

Goallessness or Goalfulness? At the present state of encounter-group culture it is almost impossible for the participants to start goalless. First of all, the sponsoring agency and the facilitators all have certain goals and processes in mind, even if these are not openly shared with the participants before the group begins. Second, even first-time participants have certain goals in mind, for they have talked with friends who have been in groups or they have read about encounter groups in the press or in a variety of popular books now available. Therefore, to picture the group as it begins as goalless is unrealistic. Goals, however hidden, abound! If this is the case, then it is more reasonable to make some sense out of all this goal confusion before the group begins.

Goals in Encounter Groups

Rather than planned ambiguity of goals and processes, I believe in high visibility and clarity. The prospective member of an encounter group should know what is expected of him.

The overriding goal of a laboratory in interpersonal relations is, obviously, interpersonal growth. But interpersonal growth is far too general a term and must be defined operationally. I grow interpersonally if

- I am freer to be myself in my interactions with others,

- I manage interpersonal anxiety more effectively,

- I learn how to show greater concern for others,

- I can take initiative in contacting others more easily,

- I can share myself more openly and deeply with the significant others in my life,

- I can be less fearful in expressing feelings and emotions in interpersonal situations,

- I can step from behind my facade more often,

- I can learn to accept myself and deal with my deficits in the community of my friends more often,

- intimacy frightens me less,

- I can endure concerned and responsible confrontation more,

- I can learn to confront those who mean something to me with care and compassion,

- I can come to expect myself and others to work on the phoniness in their lives,

- I can commit myself more deeply to others without fear of losing my own identity,

- I can come to know who I am a bit more in terms of my personal goals and the direction of life.

This, admittedly, is a large order. Involvement in a short-term group may be only a beginning, but it can be a good beginning.

Contracts as Ways of Operationalizing Group Goals. If goal specificity is desirable in encounter groups and in other situations in which the small group is used as a vehicle of personal and interpersonal growth, then the notion of contract can be offered as a possible way of clearing up some of the confusion associated with both the conceptualization and practice of these groups.

A formal definition of contract is "an agreement enforceable at law made between two or more persons by which rights are acquired by one or both to acts or forebearances on the part of the other." Justice Holmes has remarked more cynically that a contract is the "taking of a risk." The notion of contract, as it applies to the present discussion, probably stands somewhere between these two definitions. A common-sense understanding of contract is all that is necessary for our purposes and neither legal niceties nor cynicism will add much to the discussion. Contracts need not be legalistic. They can be quite humanistic. Indeed, they are what we make them—marriage is both a beautiful and dismal witness to that.

A contract to which prospective participants subscribe *before* they enter the group experience can help stimulate a high degree of operationality in the group in various ways. First of all, it defines the group experience and sets it apart from other kinds of small-group process. The prospective participant, then, knows in general what he is getting into. Not only good ethics but the logic of commitment seems to demand that participants know what kind of group they are about to join. It is hypothesized here that such clarity of commitment would ensure a higher degree of what can be called psychological (participative) rather than merely formal (spectator, nonparticipative) membership in the group. Second, a contract, in a practical way, makes high visibility rather than traditional encounter-group ambiguity a reality for the group. High visibility means that the members know what they are getting into. Both the goals and the procedures for attaining those goals are highly visible. Third, a contract outlines the procedures and processes of the group; that is, it links means with ends. A contract, then, by clarifying the demands made on the participant and the facilitator, can help provide a degree of psychological safety that is unfortunately sometimes lacking in encounter groups.

A good contract is not meant to control members, to restrict their freedom unduly. Its purpose is rather to channel the energies of the group toward specific goals. The contract should leave each participant a good deal of room in which to move. For instance, while the contract indicates that self-disclosure is a value, it does not dictate either the content or the level of self-disclosure. Each participant must determine for himself how and to what degree he is going to reveal himself in the group.

One of the needs a participant has as he goes through the group experience is to know whether he is succeeding. But without explicit group and/or personal goals, this knowledge is difficult if not impossible to obtain. A contract can provide the participants with some kind of concrete criterion for judging achievement. A sample contract addressed to the prospective encounter-group participant is presented below. It outlines the principal goals of the group and the processes that research has shown necessary to achieve these goals.

An Encounter-Group Contract

Goals

The Overriding Goal. The most general goal of the group experience is interpersonal growth. But such a goal is *too* general. The whole function of this contract is to make this goal more concrete and to spell out the kinds of group processes that lead to this goal.

The General Procedural Goal. A general procedural goal is this: you are to work for the establishment of an intimate community within which the members support and cooperate with one another to the degree that each of you feels free to investigate your interpersonal style and to experiment with interpersonal behaviors that are not normally part of that style.

A More Specific Procedural Goal. This goal is simple to state but difficult to put into practice: each member of this group is to try to establish and develop a relationship of some closeness or intimacy with each of the other members of the group. Each member should try to come to know each other member in more than just a superficial way. This goal is difficult to put into practice because it means that each person must take the initiative to go out of himself and contact each of the other members of the group. You will probably not be successful in establishing a relationship of some closeness in each case. However, you should learn a great deal about yourself as you observe yourself engaging in this process and as you receive feedback concerning your interpersonal style from the other members of the group. You can learn from both your successes and your failures.

Assessment as a Goal. As you interact with the other members, you will both observe your own behavior and receive feedback from your fellows with respect to the impact you are having on them. This feedback will give you an opportunity to get a clearer picture of and deeper feeling for your interactional style. Group interaction gives you the opportunity to assess your areas of strength and your areas of deficit or weakness in interpersonal relating.

Experimentation with "New" Behavior as a Goal. As you learn more about how effective or ineffective you are in contacting others—by observing both your own strengths and weaknesses and those of others—you can attempt and practice the kinds of behavior that bring about more creative contact with others and change the behaviors that prevent this. This is what is meant by "new" behavior. For instance, if by watching yourself and others and by receiving feedback you come to realize that you do not listen well (you are not effectively receiving the verbal and nonverbal messages of others and responding to them), then you can concentrate on listening more effectively—that is, more actively. In your group there might well be people who listen quite well. The good listeners, then, become models whom you can imitate (just as others might learn from the things you do well).

Personal Goals. You might have personal goals, personal reasons for engaging in the group (perhaps to learn about group processes). Your personal goals might be identical to those outlined in this contract or they may be different. Your personal goals and the ways they might conflict with the contractual goals of the group should be shared openly with the other participants, for the group will stagnate if individual members pursue their own "hidden agendas."

Interactions

Certain interactions are common to all encounter groups because they are the kinds of interactions that put people in more human contact with one another. One function of this contract is to point out these interactional "values." If you commit yourself to these values, then the chance of establishing the cooperative community mentioned above is heightened considerably.

Self-Disclosure. Self-disclosure in the encounter group is important, but it is not an end in itself. If I want you to get to know me, then I must reveal myself to you in some way. Therefore, you should be open, but primarily about what is happening to you as you go about the work of contacting the other members of the group and trying to establish some kind of relationship with them. "Secret dropping" may be sensational, but it is not a value in the group. You are important, not your secrets. When you do reveal what is happening in your life outside the group, you should do so because it helps you to establish and develop relationships within the group and because it is relevant to what is happening here-and-now in the group. You must choose what you want to disclose about yourself in keeping with the goals of the group.

Expression of Feeling. The contract calls for expression of feelings and emotions. This does not mean that you must manufacture emotion. Rather you are asked not to suppress the feelings that naturally arise in the give-and-take of the group but to deal with them as openly as possible. Suppressed emotion tends eventually either to explode and overwhelm the other or to dribble out in a variety of unproductive ways.

Empathy. Try to see the world through the eyes of your fellow members. Listen to them actively—that is, to all the cues or "messages" they emit, whether they be verbal or nonverbal. Try to understand the experience of others, especially when it differs from your own. Finally, try to communicate to others your understanding of them. The less prejudicial this understanding is, the deeper your contact with the other will be.

Support.   The contract calls for support, whatever name you might give it—respect, nonpossessive warmth, acceptance, love, care, concern—or a combination of all of these. You will be effective interpersonally to the degree that you can actively *express* to others that you are "for" them just because they are fellow human beings. Without a climate of support, encounter groups can degenerate into the destructive caricature often described in the popular press. On the other hand, if you receive adequate support in the group, then you can usually tolerate a good deal of strong interaction. Without a climate of support there can be no trust. Without trust there can be no intimate community. Support can be expressed many different ways, both verbally and nonverbally, but it must be *expressed* if it is to have an impact on the other. Support that stays locked up inside you is no support at all.

Confrontation.   If there is an adequate climate of support, of "being for" one another, then you can benefit greatly by learning how to challenge one another effectively. Confrontation does not mean "telling the other off." This is merely punishment, and punishment is rarely growthful. When you confront, follow these two simple rules: (1) Confront only if you care about the other and your confrontation is a visible sign of that care. (2) Confront in order to get involved with the other, as one way of establishing and developing a relationship with him. Remember, it is possible to confront another with respect to his unused strengths as well as his demonstrated weaknesses. There is evidence that the former is a more growthful process. Remember too that your confrontation will be well received to the degree that you first build up a base of support for the other.

Self-Exploration as Response to Confrontation.   When confronted, try to react by exploring your behavior in the context of the encounter community instead of defending yourself or attacking your confronter. "What you say disturbs me, but I think that I should explore it with you and the others here" is not an easy response, but it can be very growthful. Both the one confronting and the one being confronted should learn to check out the substance of the confrontation with the other members of the group.

Procedural Rules

Certain procedural rules help create a climate of greater contact and immediacy in the group. The following rules, then, govern the interaction:

1. Initiative. Don't merely react to others; don't wait to be contacted by others. Take the initiative, reach out, contact others. The importance of initiative cannot be overstressed.

2. Genuineness. Be yourself. Don't be phony, don't hide behind roles and facades. Don't play interpersonal games with the members of the group. Whatever you do, do it genuinely and sincerely. This will help a great deal to establish a climate of trust in the group.

3. Concreteness. Be direct, concrete, and specific in your interactions. Avoid speaking about generalities and abstractions and theory. Speak about behavior—yours and that of other members.

4. Speak to Individuals. As a general rule, speak to individual members rather than to the entire group. After all, a major goal is to establish and develop relationships with individual members. Speeches to the entire group do not contribute effectively to this end. Furthermore, such speeches tend to be too long, boring, and abstract. The group cursed with consecutive monologues is in bad straits. Aim at spontaneous dialogue.

5. "Own" the Interactions of Others. Part of taking initiative is "owning" the interactions of others. In the group, when two people speak to each other, it is not just a private interaction. Other participants may and should "own" the interaction, not just by listening passively but by contributing their own thoughts and feelings when they are deemed appropriate. Your tendency will probably be not to own the interactions of others because you

do not want to "interrupt." Being tactless is one thing; spontaneous involvement, motivated by care, is another.

6. *Speak for Yourself.* Avoid using the word "we." When you
use "we," you speak for the group, and you should, generally,
speak for yourself. The word "we" tends to polarize; it sets the
person spoken to off from the group. "I don't think you are really
with us this evening" should be "I miss you, John; I'm bothered
by your silence." Furthermore, when you are speaking for yourself, use the pronoun "I" rather than less immediate and distancing substitutes—"we," "you," "one," "people." Strangely
enough, the pronouns you use can make a difference.

7. *Say It in the Group.* A wise person once said that there
is an excellent criterion for determining the level of trust in a
group: Do the members say inside the group the things they say
outside the group (to one another privately, to wives, to friends)?
As much as possible, then, say what you mean in the group.

8. *The Here-and-Now.* Deal with the here-and-now. When you
talk about things that are happening or have happened in the past
outside the group, do so only if what you are saying can be made
relevant to your interaction with *these* people in *this* group. This
does not mean that you should never deal with the there-and-then,
but do so in order to further the task of establishing and developing
relationships within the group. The there-and-then can prove
quite boring even when it takes the form of analyzing past interactions of the group.

Leadership

The facilitator is in the group because he is interested in interpersonal growth. While it is true that he brings certain special
resources to the group because of his theoretical background and
experience, his purpose is to put whatever resources he has at
the service of the group. He subscribes to the same contract as
the other members do. In the beginning the facilitator will be
more active, for one of his functions is to model the kinds of
behavior called for by the contract. Another way of putting this
is that he will strive to be a good member from the beginning.
Another one of his functions is to invite others to engage in contractual behavior. However, the ideal is that whatever leadership

(in terms of contractual behavior) he manifests becomes diffused in the group. Eventually in the group there should be no leader but a high degree of shared leadership. This will be the case if individual members take the initiative to contact one another according to the terms of this contract.

Expect the facilitator to be a good member, but don't put the responsibility for the well-functioning of the group on his shoulders. Be active, take initiative, seize the responsibility for making the group a good one yourself. If you don't care for the facilitator's style of leadership, talk it out with him.

There is nothing magical about a contract; they are only as effective as the willingness of group members to participate in and fulfill them.

The rest of this book is an attempt to put some flesh and blood on the bare bones of the contract described here, to explain briefly why specific elements have been included, and to give examples of good group interaction.

Chapter 4

Leadership

Leadership is an important topic in laboratory groups, for it does not deal merely with the tasks of the appointed leader, the facilitator; rather it deals with the behavior of each member in the group. To put this in a slightly different way: all laboratory training is, in some sense, leadership training.

Over the past sixty or seventy years a great deal of research has taken place in the area of leadership. For a long time researchers barked up the wrong tree, for they studied leadership almost exclusively as a prerogative of a particular person. They studied especially the kinds of traits a leader should have—intelligence, adjustment, assertiveness, warmth, and so on. But this trait approach to leadership proved rather sterile. There is a tendency today to view leadership as a social-influence process and not as a fixed state of being.

Leadership in a very fundamental sense refers more to the needs and goals of a particular organization or community (such as the community that is the encounter group) than to the traits of designated leaders. If the needs of any given community are fulfilled, then the community will move closer to its goals. *Whoever* contributes to the fulfillment of the needs of the community participates, by that very fact, in the leadership function of the community. For instance, support is one of the needs of the encounter-group community. Whoever effectively provides support participates in the leadership function of the group. On the other hand, whoever speaks with a great deal of cynicism in the group participates in a negative way in the leadership function of the group, for he contributes to the creation of a climate of distrust that prevents the group from moving toward its goals. Leadership, then, is a complex process of social influence—it refers to goals, the nature of the group, the needs of the group, the nature of the designated leader and his functions, the various situations in which the group finds itself, the needs and the characteristics of the members of the group, and other variables.

It is both strange and unfortunate that much of the writing about leadership in encounter groups ignores the research literature on leadership. This has simply intensified the confusion of a field that is already overly confused.

The Leader in the Encounter Group

Most encounter groups meet under the direction (or nondirection) of a leader; the encounter-group movement helps him escape the odium associated with the rubric "leader" by calling him something else—trainer, educator, facilitator. In the following pages the word "facilitator" will be used, since in a negative vein the word "trainer" can connote manipulation (animals have trainers) and the word "educator" is too closely linked with a process that arouses negative feelings in many. In a more positive vein, a "facilitator" is one who offers whatever services he can to help others develop their own resources and talents.

Not all groups meet with a facilitator. Gibb (1964) has experimented for years with groups with no designated leader. In view of the present consternation over the functions and qualifications of facilitators in encounter groups, Gibb's remarks are quite challenging:

> Our many years of experience with "leaderless" groups in various settings lead us to feel that maximum participative behavior is attained more readily in training groups without trainers than with trainers. The groups are perhaps more aptly described as "leaderful," in that what occurs is not an abolition of leadership but a distribution of leadership roles in the group. It is perhaps more accurate to describe the participative groups as "trainerless." Members learn to observe and experiment upon their own behavior in increasingly creative ways. They learn that it is less adaptive to take a "trainer stance," that is, advise, "help," teach, change, or persuade others [pp. 298–299].

Others think that facilitators are absolutely essential for groups in order to help keep the group out of danger and to provide alternatives when the group reaches an impasse. Still others say that facilitators are necessary if the group is goalless and unstructured but that well-structured experiences do not need leaders. One group of professionals has provided a set of tape recordings to provide structure for leaderless groups. Against this confusing background of differing opinions with regard to the leader and the leadership process in the encounter group, I shall outline here my own approach to leadership.

Since I believe that more is generally accomplished in groups if the goals of the group are clear and if there is some agreement among the members with respect to the ways of pursuing these goals, my remarks are based on these assumptions. Encounter-group members can, from the outset, commit themselves to the basic structure of the group. This basic structure, as we have seen in the previous chapter, can be called a contract. It is within the structure of this contract, whether it is expressed or implied (in most groups it is implied), that I discuss leadership.

Research has shown that the designated leader in any group will be more effective if he is adjusted and congruent—that is, if he is a person who responds as the real person he actually is, who employs no artificial front and does not have to hide or fear his real reactions. The facilitator in the encounter group should be adjusted and congruent, but this does not put him in a special category or give him special prerogatives. It is not essential that he be the best adjusted member of the group or the most congruent. The facilitator-member is in the group because he is interested in interpersonal growth, his own included. He is there not because he has "made it" in the area of interpersonal relations but because he thinks that interpersonal growth is important. Because of his experience, he may be more aware of his own interpersonal strengths and deficits than many others. Ideally, the facilitator-member is a person of high social intelligence; that is, he has a feeling for people and knows how to make deeper contact with others without manipulating them. If the facilitator is not congruent and socially intelligent in other respects, then his presence in the group will be disturbing rather than facilitating, and the participants will have to spend a good deal of energy learning how to deal with him.

I see the facilitator as a member—a member with special resources—who is pursuing the same goals in the same way as the other members of the group. Not all facilitators see themselves in the same way, however. Some facilitators, at least initially if not throughout the entire life of the group, play a decidedly non-directive role. They seem to be neither leaders nor members, for they give little direction and in general interact little with the other participants. Other facilitators assume an opposite role. They are the gurus, the priest-facilitators. They take charge of the group process by the force of their personalities. They have been likened to conductors of symphonies. In one group a facilitator of this stamp was berating a priest for being too "priesty" in the group. The humor of the situation did not escape one of the participants as she thought to herself: "If some facilitators are priests, this guy is a cardinal." I have misgivings about the leader-priest model. The need for good encounter-group experiences is too great, and there are just not enough charismatic leaders to go around. Participating in a group with a challenging, exciting maestro can be a beguiling, engaging, rewarding experience, but such maestros are comparatively rare, as rare as really excellent teachers. If we wait around to be educated by inspirational heroes of education and "saved" by guru-facilitators, we will probably remain both ignorant and lost. The overly active facilitator stifles the initiative of the participants, and initiative, at least in my thinking, is central to the encounter-group process. I do not much care for the overly nondirective facilitator either. He remains a tantalizing mystery and therefore absorbs too much of the energy of the group. He too easily becomes the target of the projections of the members.

As soon as *anyone*, whether very active, moderately active, or passive, is placed in a group as some kind of leader, he becomes a force to be reckoned with. Independent members might choose to ignore him or deal with him as they would with any other member, while the more dependent members might engage in various forms of dependency or counterdependency behavior. A contract can handle this problem in two different ways. (1) It can provide sufficient structure so that a group leader is not necessary. The participants become the major resource for one another and their pooled social intelligence provides the basis for a profitable experience. (2) In groups that have a facilitator—and this refers to most groups as we know them now—a contract can handle the leader problem by outlining his functions for the members of the group. If the leader has high visibility in the group, if the members have some idea of what his functions are before the group begins, and if the leader does not go beyond his legitimate functions, then he can prove to be a valuable resource in groups if only because he is a good member.

If the function of the facilitator is spelled out in the statement of goals or in some kind of contract, he will less likely tend to usurp a disproportionate amount of the time and energy of the group. We live in an era of the democratization of leadership. It would be strange, to say the least, if encounter groups were to become bastions of elitist leaders.

The Functions of the Facilitator

Initial Structuring.   If there is to be some kind of structuring according to goals or a stated contract, such structuring should be done by the sponsoring agency before the group begins. The facilitator should be familiar, both theoretically and experientially, with the contract under which the group will be working. One of the ways in which groups tend to run away from hard work is to spend a great deal of time talking about the structure of the experience. The facilitator should take the lead in getting down to the work of the contract.

The immediate affective impact that the facilitator has on the group is relatively important. He is in a position of power, and if he acts in ways geared to arouse the antagonism of the other members, they will likely spend a great deal of time dealing with him rather than with one another. There is no reason why the facilitator cannot be warm and accepting from the very beginning, rather than aloof and ambiguous. In the encounter group, a poor beginning due to a clumsy facilitator is simply uneconomic, for it is a time-consuming undertaking to try to correct mistakes once they have been made.

Knowledge and Experience at the Service of the Group.   The facilitator should not pretend that he is a novice in groups (though he can readily admit that each new group is a unique challenge). The primary function of the facilitator is to place all of his resources at the service of the group as directly and unambiguously as possible. He is both to fulfill the expressed or implied contract and to help other members to fulfill it. He is a kind of social engineer who is interested in the development of conditions necessary for the establishment of an intimate community. His leadership does not place him outside the group, nor does it give him special prerogatives. He is not even different from other members in that he is to *serve* the group, since it is the function of all the participants to serve the group. However, his special knowledge, skills, and experience may place him in a special position to serve the group.

Because of his knowledge of group dynamics, he often knows, even before the group begins, the kinds of problems that will most likely arise naturally and impede the progress of the group. He is in a unique position, then, to help the group avoid wasting its time or at least to learn from mistakes made. It takes a good deal of sensitivity and care to be able to do this without manipulating others or treating them as children.

The Facilitator as Model. One of the best things a facilitator can do is to engage in the kinds of behavior that make for good group interaction: he accepts, encourages, engages in relevant self-disclosure, invites others to self-examination, responds to confrontation by examining his own behavior in the community of the group, expresses his feelings, cooperates with others, sticks to the here and now, owns the interactions that take place between others, tries to involve himself with others, encourages others to involve themselves with him, and generally searches for new ways of being present to others. Modeling demands a good deal of tact. For instance, if the trainer engages in self-disclosure, he does not rush in with a degree of self-revelation that would shock and inhibit rather than challenge and encourage. He gears his disclosures to the task of establishing and developing relationships with the other members of the group rather than to drama.

The Facilitator as Guardian. If the group has goals, the facilitator tries to see to it that the group pursues these goals.

If there is a contract, he urges its fulfillment in appropriate ways. This means that he will have to confront others, but this is one of his legitimate functions. He also has a second, "guardian" role. No group can long exist without a climate of psychological safety. The facilitator should promote and protect this climate. In other words, he does not allow the group to claw at any particular member, and he makes sure that there is adequate support for the interactions that are taking place.

The Diffusion of Leadership.   Ideally, the facilitator becomes less and less a leader in the group and more and more a member as time passes. The functions described above are not functions of a leader but rather leadership functions in which *all* the members of the group should participate more fully as the group moves on. This means that the facilitator must be willing to share functions that were more appropriately his at the beginning of the group. It means, concomitantly, that the members of the group are willing to express more and more initiative. The diffusion of leadership is a primary characteristic of the mature group.

A Final Note.   In a way this entire book is a statement not of the functions of the leader but of the leadership needs found in encounter groups. If the sponsoring agency (for example, those responsible for running a university course which uses the encounter group as part of its process) has the participants learn something about the anatomy of groups, the agency in effect "blows the cover" of the facilitator. He does not remain the shadowy figure of the nondirective approach or the "father" or "priest" of the charismatic approach. If members are to have difficulty with him, he wants it to be because he is another group member and not because he is leader. Group members have been divided into three classes according to how they act toward the facilitator. There are the dependents, who look to the facilitator for cues; the counterdependents, who solve their dependency needs by opposing the facilitator; and the independents, who are not overly threatened by the prospect of intimacy, even with one who has been designated leader. Authority and fatherhood problems do exist, and they will be stronger in some participants than others. No statement of goals, contract, or analysis of the problems of leadership can legislate them out of existence. But if the functions of the facilitator are spelled out clearly—legitimated, as it were, by a statement of goals or a contract—such a statement paves the way to open dialogue with the facilitator on the part of the person who is sensitive to authority. Finally, if the facilitator is not doing a good job as a leader or as a member (for example, if he asks others to do things that he himself does not do), then he should be challenged and others should take over leadership functions.

Chapter 5

Self-Disclosure

One of the principal interactional goals in the encounter group is self-disclosure. Since self-disclosure of some degree constitutes an integral part of almost any kind of laboratory experience and does so in a special way in laboratories in interpersonal relations, it needs special attention, especially in view of the fact that most of us fear self-revelation to a greater or lesser extent and therefore find it difficult to estimate its value in interpersonal living.

In all groups in which interpersonal psychological growth is at least an implicit goal, a good deal of the group's activity, especially in earlier sessions, deals with formulating policy with respect to self-disclosure. "I'm not so sure how far I can go," "I am beginning to wonder whether we are starting a therapy group here," "Boy, that [referring to some disclosure made by one of the participants] was a bomb"—these are some typical statements heard as a group tries to grapple with the problem of self-disclosure. The prospect of revealing oneself is unsettling and is approached gradually in most groups. While many of the members of encounter groups realize that self-revelation, if engaged in responsibly, is a value in human living, still they need time to screw up the courage needed to talk about oneself. The kind of group being discussed in these pages—especially the group governed by a contract—does not have to grapple with the question of *whether* self-disclosure is a value in the group or not. It *is* a value. In such a group other questions are more important. What should I disclose about myself? How should I disclose myself?

What prevents me from disclosing myself? An attempt will be made to answer these questions in the following pages.

It has been said that a person who cannot love cannot reveal himself. The converse also seems true: the person who cannot reveal himself cannot love. The assumption of this book is that responsible self-revelation is essential to the establishment of an intimate community in the context of which members can develop the human skills necessary to relate to one another at a high level. The very sharing of the human condition—in its sublimity, banality, and deformity—pulls people together.

Self-Disclosure in the Context of the Group

One of the reasons why self-disclosure appears so frightening to many is that it is often depicted as an end in itself. Out of context self-disclosure *is* frightening. In a laboratory in interpersonal relations, however, it should always be kept in context; that is, it should always be related to the goals of the group and the individual goals of the participants. There is a world of difference between "secret dropping" (self-disclosure out of context) and self-revelation as one of the means used to establish and develop relationships of some closeness or intimacy with other group members (the procedural goal of the encounter group).

What to Disclose in the Encounter Group

If I am a participant in an encounter group, there are, broadly speaking, two categories or sources of self-revelation: (1) what is going on inside me during the group itself—that is, how I feel, what I think about myself and the other members of the group, and (2) my experience and behavior in the past (the "then") and my experience and behavior outside the group (the "there"). Let's take a look at each source separately.

My Experience Inside the Group. The more freely you and the other participants reveal what is going on inside you during the group meetings, the more effective your contact with one another will be. Consider the following examples:

- I'm a very short-tempered person and you are really getting under my skin.

- I like you, but I feel a bit foolish putting it that baldly.

- I've been bored this past hour, but I think I deserve my boredom because I've done nothing about it.

- I think my behavior here shows that I'm pretty self-centered.

- I'm confused. I'm not sure whether you are saying that you like me or that you dislike me.

- I'm curious about you, John, about what you are like. I can't say that that is friendship, much less love, but somehow I feel that it is something very positive.

These disclosures are attempts on the part of the participants to deal immediately with their relationships to others or to what is happening in the group. This kind of self-disclosure is extremely important in any group. No trust can be built up in a group if the participants feel that there are members who are keeping things to themselves. The silent member and the guarded one, then, hang like dead weights around the neck of the group. It is extremely difficult to risk trusting the silent or the overly guarded member.

Past Behavior and Behavior Outside the Group: The "There-and-Then." I can also talk about my past experience and behavior (whether inside the group or outside) and what I do and experience outside the group. Since the purpose of self-disclosure in encounter groups is not merely to retail information about myself, disclosure of there-and-then experience and behavior should be related to the purpose of the group (establishing relationships with others, increasing interpersonal skills). Here are some examples of how there-and-then material is related to the group both poorly and effectively:

Unrelated self-disclosure	*Related self-disclosure*
My dad and I don't get along. He's usually down on my brother, too. It makes living	I'm reacting to you just the way I do to my dad. He doesn't listen to me, nor do

at home difficult. Sometimes I just feel like getting an apartment on my own.

you. I feel like turning my back on you, like I feel like getting my own apartment. You're older, but I don't think that has anything to do with it.

I'm homosexual. I don't think society is fair to the homosexual. But then society is uptight about a lot of things. People have to fit in, to do things like everybody else.

I think I trust you people enough to tell you that I'm homosexual. The reason I'm telling you is not to have you counsel me. I feel that some of you might reject me, but I'm hoping you won't. Since my sexual identification is something big on my mind and since I fear rejection, I felt I had to bring it up.

I don't communicate much with my wife and children but they let me alone.

I don't communicate much with my wife and children at home, and I'm falling into the same pattern here, except you challenge me whereas they don't. I should begin to challenge myself.

I have many fears. I'm afraid of strong men. I'm afraid of those who don't like me. I'm afraid I will not succeed.

I have many fears. For instance, I'm afraid of strong men, Bill, and I see you as strong. I fear rejection, and, Jane, I feel you don't like me. I'm afraid of failure and I think I am making a mess of this group experience. Well, it's out in the open now.

What, then, should you disclose in the encounter group? The answer is simple: anything that helps you and the other members pursue the goals of the group. In other words, self-disclosure should be organic—related to the goals of the group. The depth of a person's self disclosure should be related to the process of establishing and developing relationships and should arise naturally from the give-and-take of the group process. As indicated in one of the examples above, I have seen participants reveal something as sensitive as their homosexuality *not* because they wanted to engage in secret dropping but because they needed to know that they would not be rejected in the group because of their sexual identification. They did not spend a great deal of time subsequently dealing with their homosexuality, for that was not a here-and-now issue for the group. But dealing with it briefly improved their ability to communicate with *these* people in *this* situation. Once this obstacle was overcome, they felt free to go about the business of developing relationships with other participants.

You, the individual participant, are in charge of your own disclosure. You should take the initiative to reveal yourself at whatever level you think you should in order to pursue the goals of the group. While you should resist any individual or group effort to claw your secrets out of you, still you should take some reasonable risks in the group. The purpose of the group is not to throw your defenses to the winds (and uncontrolled self-disclosure is one way of doing precisely that), but you should be willing to experiment with a reasonable lowering of your defenses. One way of doing this is to risk revealing yourself and your feelings if you think it will help you establish and develop relationships in the group, *even* if you fear some rejection on the part of some group members. First of all, most rejection fears in high-level encounter groups are unfounded (for there is a climate of support), and, second, you may also learn that you can live on, even though some people cannot approve of certain aspects of your personality or behavior. Most of the outstanding figures of history had to face a great deal of rejection precisely because they had the courage to tell the world who they were and where they stood.

The Quality of Self-Disclosure: "Story" versus "History"

The *way* in which a person reveals himself in the group is very important. In a sense it is even more important than the content of the revelation, for content, no matter how intimate in itself, can lose its intimacy and its meaning in the telling. I propose two styles or modes of self-disclosure: "history," the mode of noninvolvement, and "story," the mode of involvement.

History. A recent television documentary showed excerpts

from a marathon group experience conducted at Daytop Village, a rehabilitation center for addicts. During the early hours of the marathon a young addict began talking about himself and his past life. His self-revelation was almost totally history. I was disturbed to think that what he was doing was considered acceptable group process; that is, I was disturbed until one of the group leaders finally spoke up and confronted the speaker. In effect he said: "You have been engaging in history rather than story, and mere history in this group experience is meaningless."

History is pseudo-self-disclosure. It is actuarial and analytic, and usually has a strong "there and then" flavor. It clicks off the facts of experience and even interpretations of this experience but leaves the person of the revealer relatively untouched; he is accounted for and analyzed, but unrevealed. The person relates many facts about himself, but the person within still remains unknown. History is often a long account. It is long and often steady because it fears interruption. Interruption might mean involvement, and a person engages in history to avoid, rather than invite, involvement. History has a way of saying "Be quiet" or "Don't interrupt," but these are dodges to keep others at bay. The steady clicking off of facts keeps the group focused on the revealer but does not allow the members to involve themselves with him.

In history the manner of self-revelation is usually somewhat detached. There is little ego-involvement and thus little risk. The speaker deals with himself as object rather than as subject. Intimate life details might be revealed, but the intimacy has no particular meaning. The details are just facts. On the other hand, history might be a string of generalities—generalities poorly disguised by the first-person pronoun. But whether history consists of intimate details or generalities, the message is always the same: "Keep your distance." It is as if the revealer were trying to intimate to others that he is rather invulnerable: "This is not really affecting me; I don't see why it should affect you." Sometimes sheer quantity of intimate information about self is divulged because the historian implicitly realizes that if he relates enough, quickly enough, the others will not be able to react effectively to any particular part of it. History is also self-centered. The leader in the Daytop Village marathon took the young addict to task for his egocentricity. He told him that he had been talking a long time and had not even mentioned that he had a wife who had feelings.

Historical information does not unite speaker and listeners. Rather, the information sits there as an obstacle between them. It is a barrier rather than a bridge. It sometimes even has an "I dare you to do anything about this" aura. Even when the information disclosed is intimate, it is usually boring. The historian exudes an "I don't really care" attitude that is readily picked up by the other members of the group. The information *is* boring, because it is divorced from the person. It is flat; there is no human drama about it. History is "computorial," and, as such, calls for feedback rather than human response. In McLuhan's (1964) terms, history is a "hot" modality, high in definition and low in involvement. Its high definition refers not to just sheer quantity but to its "there it all is and there is really nothing to do about it" quality.

Story.   Story lies at the other end of the continuum. It is authentic self-disclosure, for it is an attempt to reveal the person within; and more than that, it is an attempt to get him involved with his listeners. Story is an invitation for others to come in; it is an opening of the door. In group growth experiences, as in the rest of life, others often stand around waiting to come in. Story is a signal for others to move into one's presence.

Story is not actuarial; it is rather selective in detail, for the revealer intuits that it is not the transmission of fact that is important but the transmission of self. It does not avoid detail, but the choice of detail is secondary to the act of communication. Story usually avoids interpretation, too; it allows experience to remain unintellectualized and thus speak for itself. The storyteller, even if he leaves out detail, is graphic and specific, he does not hide behind generalities disguised by the first-person pronoun. Facts are selected for their impact value, for their ability to reveal the person as what he is now through what he has experienced.

The storyteller is taking a risk and he knows it. Therefore, story is always an implicit request for human support. The revealer has come to trust the group to a certain degree, but he still feels his vulnerability; his act of self-revelation is akin to

Kierkegaard's "leap of faith," which is always a leap of trust. But he takes this leap because he wants to relate to the other members of the group and relate more fully to himself. He realizes that story is the way of involvement and the way of discovery, and he wants both. And so he comes to the point. He does not wander around in the there and then but manages to make the past, the "there," and even the future define him as he is in the here and now. Story, then, is not analytical and discrete. It is synthetic; it attempts to present a totality, the complex totality that is the person himself, who takes shape out of the complexity of his experience.

The one who tells his story—if that story is not computorial and therefore not a request for feedback in a dehumanized sense —is looking for human response. Story, of its very nature, is dialogue and merits such response. Because story is not computorial and monologic, it is inevitably engaging, even when someone who is usually a bore adopts it. Some people are constantly talking about themselves, and most others find this terribly boring. Such people are boring because they are usually engaging in history rather than story. First of all, they are really saying nothing about themselves, and second, they care little for the objects of their monologue and would find real response, such as self-disclosure and confrontation, frightening. Bores speak in generalities poorly disguised under the pronoun "I." But story, on the other hand, is always engaging, for it means that the speaker has to "blow his cover," lower his defenses, and stand somewhat naked in his own eyes and in the eyes of others. People are seldom, if ever, bored with sincere self-revelation, because they intuitively realize its importance for the one revealing himself and respect him for what he is doing. The person who engages in story is one who stops complaining about how much he *hurts* and begins admitting who he *is*. I think perhaps that it might be impossible to dislike someone who engages in story, for it is an act of humility, a manifestation of a need to move into community, and a surrender of egocentricity (or at least a beginning of surrender).

In McLuhan's terminology, story is a "cool" modality, low in definition and high in involvement. It is low in definition not just because it is selective of detail and thus allows others to fill in the gaps in information, but because the information transmitted is seen as a medium, a bridge instead of a barrier. Story has high impact value; it tends to change both speaker and listener. It draws the listener out of himself and toward the speaker; it changes the speaker in that it calls forth emotions that are authentic and therefore perhaps unfamiliar to the revealer. Story, then, is not maudlin, but it is shot through with emotion; it is not sensational, but it has drama in the same way that a life fully lived has drama.

Self-Disclosure in a Wider Context

To consider self-disclosure only as a dimension of an encounter-group experience would be to view it too narrowly. To understand the place of self-revelation in the encounter group, let's see how it fits—or does not fit—into life.

Deception and Concealment as Growth-Stifling

A number of psychologists blame deception and concealment for many of the emotional problems we suffer. O. H. Mowrer (1968a, 1968b) and Sidney Jourard (1964, 1968, 1971) are the principal proponents of this position and both have amassed data to support it. Mowrer's basic thesis is that men spend too much energy trying to conceal both from themselves and others the fact that they engage in various forms of irresponsible behavior. In one model, he divides the usual symptoms associated with emotional disturbance into two types. Type I symptoms are the agonies usually associated with emotional distress of one kind or another: tension, anxiety, depression, fatigue, loss of appetite, loneliness, phobias, scrupulosity, a sense of unreality, hypochondriasis, and so on. He sees these symptoms, at least in many cases, as discomforts arising naturally from deviant behavior. When someone acts irresponsibly he *should* feel bad. These symptoms are not under the sufferer's control; they are mediated by the autonomic nervous system. Still they are potentially useful (as pain is useful) to motivate the sufferer to change his life style.

Type II symptoms are the means, usually ineffectual, that the sufferer uses to try to handle Type I symptoms. They are attempts to escape the pain rather than attempts to get at its roots. These symptoms include withdrawal, suicide, rationalization, blaming others, self-pity, busyness, overeating, abuse of sex, daydreaming, intoxicants, and tranquilizers. Type II symptoms prevent the sufferer from moving in the right direction. They are "home remedies" that generally do not work. Since they are delaying tactics, they ultimately make things worse. Both Types I and II symptoms are muted confessions of self-defeating behavior or even self-defeating life styles. Mowrer contends that many people could rid themselves of these symptoms (both I and II) by "confessing" or revealing one's life style completely, at least to the "significant others" in one's life. Put succinctly, sometimes emotional disturbance loses its power when it loses its privacy.

Mowrer is not blind to the fact that some disorders may well have a biochemical rather than a behavioral basis. But he sees great danger in the overextension of the biochemical or so-called medical model of emotional disturbance. He readily admits that to the extent that emotional suffering has a biochemical basis, chemical and medical intervention is a boon to humanity. But when suffering is the consequence of behavioral malfunctioning, then the sufferer has re-educational work to do, and chemicals will not really solve the problem. Mowrer's model, then, refers to problems with a behavioral and interpersonal rather than a strictly organic, biochemical basis.

Jourard has been investigating the consequences of concealment and self-disclosure in a context that is wider and perhaps more positive (and less "moralistic") than Mowrer's. Like Mowrer, Jourard believes that concealment can sicken a person. The person who finds his behavior unacceptable, in one way or another, both to himself and to others, must conceal his own identity. The energy that he pours into concealment adds to his stress and dulls his awareness of his own inner experience. Whatever contact he makes with others is through a facade; a kind of rigidity or stereotypy permeates his relationships with others. Loneliness and depression are inevitable as part of the price for concealment, for the concealer is separate, apart, out of community. The concealer thus increases the stress factors in his life and thus becomes susceptible to all sorts of sickness, both physical and psychological. But when Jourard talks like this, he is not describing just the neurotic or the emotionally disturbed. He sees this problem as the affliction of most people, at least in our own society. This lack of transparency is a major element in the "psychopathology of the average" that afflicts the so-called "normal" personality of our time. The more desperate the need to conceal, the greater the stress, and the more likely the occurrence of physical and psychological distress.

Perhaps neither Mowrer nor Jourard definitively proves that concealment in itself causes emotional disturbance or that self-revelation, even when accompanied by appropriate behavioral change, effects a cure. However, both offer much empirical research concerning the negative effects of guilt- and shame-provoked concealment as well as the positive, self-actualizing effects of self-revelation.

Cultural Bias against Self-Disclosure

The society we live in tends to ban intimate self-disclosure for a number of reasons. It is worthwhile to examine a few of them here.

Self-Disclosure Seen as a Weakness.   The person who exhibits strength by suffering in silence has become a cultural stereotype in our society. "Little boys don't cry" is an early version of the masculine ideal, and the woman who, in the fiction of radio, TV, or the novel, confesses that "I simply have to talk to someone" is thought to be really confessing not a deep human need but her own weakness, even though such weakness might be understandable and even excused in a woman. If self-disclosure is not weakness, then it is seen as exhibitionism and, as such, is a sign of illness rather than a desire for human communication. Very often the adolescent, in his discovery of himself and the "other," engages in a good deal of self-disclosure. But this drive to exchange intimacies, even though it might have overtones, at times, of exhibitionism and other kinds of problem behavior, should be looked upon as the beginning of something that could be quite good—being at ease in discussing oneself with significant others at an intimate level. Instead, it is often considered adolescent behavior, naive and immature. Such behavior, it is thought, will pass, just as the "natural neurosis" of adolescence passes.

When the adult finds it necessary to communicate himself at an intimate level to a friend, he often feels that he needs an excuse to do so. The person with a painful and perplexing personal problem is often loath to ask a friend to share it, and his friend is loath to encourage him to talk about it. It is difficult for both of them, for there is little cultural support for what they are doing.

The Medical Model and Cultural Permission. Even a society that is somewhat afraid of honesty cannot ban self-disclosure completely. One person's communicating himself intimately to another, especially in times of special stress, is such a basic need that even a relatively closed society must find ways of channeling such disclosure, must find cultural justification for it. Freud was a courageous man. He took a bold step forward when he declared that the revealing of intimacies about oneself was a medical act and, as such, was perfectly justified in any society. Society could hardly be accused of greeting Freud's thesis with wholehearted approval. Still, over the years, it has become quite acceptable to reveal oneself to a doctor or to a psychologist or counselor. Intimate self-disclosure becomes justifiable in such a context. Society's way of allowing intimate self-revelation is also its way of containing it: it should take place between a client and a professional. Seeking therapy or counseling gives a person the cultural excuse he needs to establish a relationship in which he is free to tell anything about himself. But "therapy" implies illness, and even "counseling" implies problems, so that, to a large extent, self-disclosure is still associated with weakness.

Group therapy widened the scope of cultural permission to reveal oneself. One was now allowed to reveal himself to peers, provided that some professional presided over the group interaction. The present popularity of all sorts of "growth" groups, encounter groups included, is another breach in the wall. People do not want to declare themselves ill in order to engage in intimate communication about themselves.

Our Overemphasis on the Value of Privacy. Much has been written about the value of, and the individual's right to, privacy. But the coin of privacy has two sides, and Bennett (1967) takes a rather refreshing look at the obverse side:

Privacy ... is a graceful amenity, generally to be fostered, but
with discriminating restraint and with due recognition of obligations as well as privilege. It is the writer's contention that the
moral imperative is more often allied with the surrender of privacy than with its protection.... Secrecy within the community
is incompatible with cooperation, inimical to the welfare and
progress of the ingroup.... Strictly speaking, of course, sex is
not ordinarily a private experience, but a peculiarly delicate and
intimate transaction between at least two people. I submit that
even in this sensitive area, more serious problems stem from mis-
managed communication about sex—partners who cannot discuss it, children who must not be told, and alienation of the
deviate—than from mere breaches of privacy.... The confessional is also respected as a confidential relationship. It should
be noted that this, too, is a communication; a revelation, in fact,
of the most private secrets to at least one other person.... The
reference, in many religions, is to public confession. The Protes-
tant sinner must bear witness "before men" to achieve absolution.
Indeed, it is recognized in Catholic circles that the traditional
confessional, intent on making peace with God, leaves
unresolved the problem of making peace with the community....
The readiness of people to discuss their personal problems with
neighbors, and even strangers, makes one wonder, in fact,
whether confidentiality is so necessary to the privacy of the
patient as the comfort of the therapist. There are therapists who
believe that the therapeutic process is facilitated in the presence
of an audience. The popularity of group therapy reflects a similar
assumption that patients find help in sharing personal problems
—that confession is good for the psyche as well as the soul....
The contemporary concern over privacy parallels a pervasive
need to communicate.... Our dilemma will not be resolved by
hiding away from each other in separate caves, but through more
and more interpersonal communication, better managed.... The
critical problem we face is not how to keep secrets from each
other but how to facilitate this readiness to communicate. The
overriding question is how to maintain an atmosphere of trust
and confidence which will enable us to talk about personal affairs
... freely.... It is the writer's conviction that the importance
of honest communication in our interdependent relationships
outweighs the sanctity of privacy as a social value.... Anyone
who undertakes to influence the lives of other people must accept
an obligation to let them know where he stands, to reveal his
motives, to share his purposes [pp. 371–376].

Perhaps many would take exception to much of what Bennett
has to say, but he does confront individuals and society in the

spirit of such men as Mowrer and Jourard. Such confrontation merits a good deal of self-exploration (and exploration of society) as a response.

Resistance to Self-Disclosure

Why do we resist even responsible self-disclosure? However conditioned we are, we can hardly blame society for everything. We seem to flee self-disclosure and self-exploration in the community of those who care for us because we fear self-knowledge, intimacy, responsibility and change, and rejection.

The Flight from Self-Knowledge. Self-disclosure is one of the principal ways of communicating not only with others but with oneself. Perhaps the latter is even logically prior. An intriguing hypothesis is that many of us flee self-revelation because we are afraid of closer contact with ourselves. It has been said that the human organism seems capable of enduring anything in the universe except a clear, complete, fully conscious view of himself as he actually is. Self-disclosure crystalizes aspects of the self that a person would rather live with silently—however painful the living—than face. At least in this respect, then, a group is only as threatening to a participant as he is to himself. Inevitably it is the individual who is his own severest judge.

Jourard (1964) speaks out strongly to this very point.

> When a man does not acknowledge to himself who, what, and how he is, he is out of touch with reality, and he will sicken and die; and no one can help him without access to the facts. And it seems to be another empirical fact that no man can come to know himself except as an outcome of disclosing himself to another person. This is the lesson we have learned in the field of psychotherapy. When a person has been able to disclose himself utterly to another person, he learns how to increase his contact with his real self, and he may then be better able to direct his destiny on the basis of knowledge of his real self [p. 5].

> When I say that self-disclosure is a symptom of personality health, what I really mean is that a person who displays many of the other characteristics that betoken healthy personality . . . will also display the ability to make himself fully known to at least

one other significant human being.... Neurotic and psychotic
symptoms might be viewed as smoke screens interposed between
the patient's real self and the gaze of the onlooker. We might call
the symptoms "devices to avoid becoming known."

A self-alienated person—one who does not disclose himself
truthfully and fully—can never love another person nor can he be
loved by the other person [p. 25].

Much is being written about alienation and about identity conflicts, with attempts being made to establish both the social conditions and the intrapersonal dynamics of these problems. Man's
flight from himself is, in large part, a flight from communication
with himself. Self-alienation is frightening, but any kind of
intimate contact with the problem self is seen as even more
frightening. Self-alienation, then, becomes self-reinforcing, its
reward lying in its being, supposedly, less painful than its alterna-
tive. Even when a person gets out of contact with himself and
with others to the degree that he flees to a mental hospital, this
is still no guarantee that he is ready to face himself. Time and
again in mental hospitals, when a patient is faced with the choice
between the pain of alienation and the pain of therapy, he chooses
the former, unable to find the courage to be. Since a similar
dynamic is seen as operative in the psychopathology of the average, self-disclosure is stressed in laboratories in interpersonal relations.

The Fear of Intimacy.   It is difficult to reveal oneself on a
deep level to another without creating, by the very act of self-
revelation, some degree of intimacy. In a group situation, for some
reason, this intimacy seems to have a special intensity. The participants in various growth-group experiences are aware of this, and
even though they might have the courage to let others see the
"mystery of iniquity" or even the "mystery of goodness" that they
are, they fear the intimacy that such an act would beget. They
do not flee self-revelation as such. They flee intimacy. For some
persons, we are told, the fear of human relations is greater than
the fear of death. The all too real possibility of intimacy frightens
many people. They prefer, therefore, to skirt self-revelation and
to avoid intimacy. Some engage in sporadic acts of pseudo-
self-revelation leading to pseudo-intimacy in a games approach

to human relationships. Others merely refuse to reveal themselves in any significant way. Their message, usually more nonverbal than verbal, is: "Don't probe." In sum, failed intimacy is a significant dimension of the psychopathology of the average. One way of counteracting this flight from intimacy is to engage in self-disclosure that is proportioned to the situation and that involves reasonable risk-taking. This is what is expected in the encounter group.

Flight from Responsibility and Change. In some cases, flight from self-disclosure is a flight from responsibility, a flight from the anxiety and work involved in constructive personal change. Self-disclosure leads to the revelation of areas of deficit and areas of aspiration in human living. It is relatively easy to avoid both these areas in day-to-day living. However, once a person declares what he finds unacceptable in himself and what goals he thinks he should be pursuing, he commits himself to change, and avoidance behavior becomes more painful. Self-disclosure commits one to conversion, to the process of restructuring one's life; it demands that a person leave the security of his own house and journey into a foreign land, and most men balk at that. If one senses that conversion is impossible, then he must avoid self-disclosure. So it is assumed that some men fear, or even deprecate, self-disclosure because of the behavioral consequences it entails. If the self-revelation takes place in a group, then the pressure to change is even greater than in a one-to-one situation, for there is the necessity of facing the pressures and demands of a community.

The Reverse Halo Effect and the Fear of Rejection. Another source of fear of self-revelation could be termed the "reverse halo effect." The halo effect refers to the fact that a person judged to be competent in one area is likely to be judged by many to be similarly competent in other areas (even when he is not). For instance, if a person is an expert in psychology, his aura of expertise tends to spread to other areas and either he or others begin to look upon his pronouncements in theology or political science with the same awe. The reverse process sometimes stifles self-disclosure in growth groups. The group member fears self-disclosure because he usually thinks first in terms of disclosing the worst in himself. If he tells the other members about incompetence in one area of living, he feels that they will assume similar incompetence or irresponsibility in related or even unrelated areas. If a person admits problems in his private life, he fears that others will assume incompetence in his professional life. This is especially true if the person's profession is closely related to human living, as is, for example, psychology. There are several ways of handling such a problem in the group. The participant can try to give a balanced view of himself, speaking alternately of strengths and weaknesses. Or the group can take up the problem of the reverse halo effect and discuss it directly. The reverse halo effect is related to the problem of stereotyping and categorizing. No one likes to be dealt with as "problem," but there is a tendency in groups to identify the participants with their problems in living, for it is easier to deal with problems than with persons.

These, then, are some of the reasons we tend to avoid self-disclosure. And yet, willingness to engage in self-exploration—a process certainly involving self-revelation—is central to the growth process. A person does not become self-actualized by accident; he must risk himself in order to grow. In one study dealing with the benefits of counseling, clients judged the therapeutic value of fifteen topics discussed during counseling. There was general agreement on the relative value of the topics. Ratings showed a high correlation between the perceived helpfulness of a topic and its disturbing qualities. The topic called "shame and guilt" was experienced as extremely upsetting, but the discussion of this area of life during the counseling sessions was considered to be most helpful.

Shame and Self-Disclosure

Shame is a much richer word and the experience of shame is, in some sense, a much richer experience than most of us have been led to believe. Lynd (1958) has written a remarkable analysis of shame and its relationship to identity. The root meaning of the word is "to uncover, to expose, to wound," and therefore it is obviously related to the process of self-disclosure. Shame is not just being painfully exposed to another; it is *primarily* an exposure of self to oneself. In shame experiences, particularly sensitive and vulnerable aspects of the self are exposed, especially to one's own eyes. Shame has the quality of suddenness. In a flash one sees his unrecognized inadequacies without being ready for this revelation of self to self; much less is he ready for exposure to the eyes of others. The external event that precipitates a shame experience might be quite trivial. A casual remark or a joke might trigger a profound feeling of shame in someone, while the person who makes the remark often remains completely unaware of what is happening inside the other. But shame could not arise, could not be touched off by "insignificant" incidents unless, deep down, one was already ashamed.

A shame experience might be defined as an acute emotional awareness of a failure *to be* in some way. It is not just a realization of failed potentiality but a painful realization of what one is not. Shame need not be related to guilt. For instance, a person can be ashamed of his own body (it lacks grace, beauty; it has grown old; it is crippled, deformed), but one's physical makeup is hardly a source of guilt. When an acute shame experience strikes, there is no defense. It floods one's being. But few recognize that such experiences have potentialities for growth. The icy clarity of self-knowledge that is central to the shame experience is usually too painful. One has to escape and forget. The wound is allowed to heal, and situations that might possibly reopen the wound are quietly avoided. But this avoidance not only constricts one's life space, it also makes one vulnerable to further shame experiences. Yet even when a person is willing, it is not easy to get at the source of shame, for this involves the demanding process of re-examining one's value system, a task that most of us avoid.

Lynd (1958) emphasizes the potential value of dealing with shame through self-disclosure:

> If, however, one can sufficiently risk uncovering oneself and sufficiently trust another person to seek means of communicating shame, the risking of exposure can be in itself an experience of release, expansion, self-revelation, a coming forward of belief in oneself, and entering into the mind and feeling of another person [p. 249].

Shame, then, can be a way of self-discovery.

Shame as Revelation of Both Self and Society.   Lynd refers to "unrecognized aspects of one's society and of the world" as playing a part in shame experiences. A young man once came to me for counseling. He had been playing basketball; a game was organized with "skins" against "shirts," and he had been asked to remove his T-shirt. When he manifested some reluctance to do so, one of the other players remarked that he was reluctant because he had a scrawny build. Even though his antagonist and the other players did not realize it, he was flooded with shame. The incident was trivial, but in an instant he realized with painful clarity that he was ashamed of his body. In the counseling session it came out that his shame went further than that. He also realized that he lived in a society that excessively extols physical grace, charm, or beauty, often to the extent that it becomes a condition for acceptance. He realized that he had more or less concurred with society in this attitude, so he was ashamed of himself for having been so ashamed of his body and he was ashamed of his society for its hierarchy of values. Now he wanted to talk about his feelings about his own body and the way that he swallowed whole the values of society. His shame could have led to either constriction or growth. He chose to have it lead to growth.

Shame Experiences in the Encounter Group.   I am in no way suggesting that shame will be the central experience in the encounter group. On the other hand, some shame experiences might be a dimension of the process of self-disclosure. The point is that shame need not be a negative experience; it does have potentiality for growth if it arises. Shame is a powerful emotion and must be evoked with some caution. I have participated in group experiences in which shame was evoked recklessly, causing a great deal of pain but very little healing. Evoking shame without providing adequate human support may be as dangerous and destructive of growth in adult life as it is in childhood. Shame involves a deep awareness of human separation; and without reunion by love, shame is sterile. A person needs a good deal of support to face what-is, but such support is far more conducive to growth than is support given to someone so that he can endure the more or less self-inflicted agonies that arise from *not* facing what-is.

Risk and Trust

Many people hesitate to disclose themselves to a group. They balance on the edge of self-disclosure as they would on the edge of a diving board. But just as the shock and pain of entering the water are short-lived and certainly outweighed by the pleasure of swimming, so the shock and pain associated with self-disclosure are mainly initial and short-lived and inevitably outweighed by the benefits of mutual sharing. Still the pain of self-disclosure and the possibility of rejection lurk around the corner, so group members bide their time.

Risk is an essential feature of growth groups. A person before risking himself will ask: Can I have faith that greater openness with these people will increase understanding, respect, love? Will increasing intimacy be an enriching rather than a corroding process? The "nothing ventured, nothing gained" truism has a special applicability to communication in laboratory groups. But even initial risk-taking demands some climate of trust. The best way of increasing the level of trust in the group is for each member to show that he is trustworthy through the way he deals with and responds to his fellows. Perhaps the notion of "kairos"—the "right moment"—has some value in laboratory groups. If the participants agree that self-disclosure is a value in the group, then each must put himself on the line at the times that are right for him. Obviously such a concept could be used to rationalize away complete failure to be open in the group, but this need not detract from its utility.

A Final Note. Self-disclosure in the group can lead to a new kind of communication or at least to a new freedom of communication. Dealing with one another is something like showing one another the houses we live in, houses in which one or more rooms must remain locked because of what is behind the doors. In the laboratory group, the participants learn the skill of opening some of these doors, and the air that sweeps through the house is refreshing. To use another simile, communication that is characterized by a fear of self-disclosure is like trying to talk to someone across a room filled with pillars. In the contract group, the participants learn how to remove some of these pillars, and, to their amazement, the house remains standing. The contract group is a training laboratory in which the participants learn the value of self-disclosure. If they learn well, then they return to the significant others in their lives with the potential for a new kind of communication.

Self-disclosure is not the only value in life, but it is a dimension of the interpersonally fruitful life. Only the individual can determine what part it is to play in his own interpersonal life.

Chapter 6

The Expression of
Feeling and Emotion

There is a twofold concern in our society with respect to emotion and its expression. First of all, there is a growing concern over the inability of many people to engage in a free, constructive expression of feeling and emotion in their interpersonal relationships. Freedom in the area of affective living has become the war cry of that challenging, intriguing, confusing, and frustrating hodgepodge of people and processes known as the "human potential movement" (see Howard's *Please Touch* [1970] for an experiential account of this movement).

On the other hand, there are those who think—and not without some justification—that feelings and emotions are being apotheosized in an irresponsible, anti-intellectual way. There is, for instance, a dizzying proliferation of group experiences that focus almost completely on feelings, emotions, touch, sensory awareness, and the like. There is an extremely religious flavor about many of these groups, and often the participants sound like cultists seeking salvation in sensation and corporal ecstasy.

The problem seems to be that of the pendulum. A society that both subtly and blatantly represses emotional expression has to expect ultimately a reaction proportioned to the degree of repression. However, there is some evidence that revolution in emotional experiencing and expressing is not as widespread as the popular press has us believe. Emotional repression is undoubtedly still a far greater problem than emotional overindulgence.

Formal education is overloaded in the area of intellect, impoverished in the area of emotion. Some critics say that schools leave emotional education to the radio with its din of popular music, to television, to the movies, and to a host of commercial exploiters with their emotionally charged but less than truthful advertisements. Pop heroes and movie stars are the schoolmasters of emotion. Popular novels, movies, and television constitute a two-edged sword in our emotional life. If used with imagination and discretion, they complement our emotional life, enhance and enrich it by broadening our emotional experience. This is likely to be the case with those who are aware of their emotional lives and interested in developing them. But many people remain, for the most part, passive victims of their emotions. For them pop music, movies, television, and popular novels are not complements to emotional living but rather substitutes for it. For too many this vicarious emotional living is sufficient.

There is a general tendency in our society to discourage emotion. The word "emotional" is used often as a derogatory term. Developing emotional maturity is more often seen in terms of training a child in what he should *not* feel and in controlling the expression of his feelings than in extending the range and depth of his emotions and their expression. No wonder, then, that the cheap sentimentality of so many popular songs, movies, and television is so compelling; it is directed toward millions of emotion-starved customers.

One way of looking at the emotional parasitism of society—that is, the fact that men become emotional parasites to television, movies, and the other things mentioned above—is that it is an essential or at least unavoidable phenomenon in a techno-cratic society such as ours. Keniston (1965) suggests that two phenomena of our society converge to create an emotional dilemma for the working man. First of all, many men find little emotional satisfaction in the work they do to earn a living. Work, instead of satisfying emotional needs, intensifies them. Breadwinners come home, then, hungering for emotional satisfaction and expecting to find it with their families. But today's family—and this is the second phenomenon—is a smaller unit than yesterday's. Family no longer means a complex of grandparents, aunts, uncles, and children living in the same relatively circumscribed geographical area. Family today means wife and two or three children, too geographically or psychologically separated from close relatives to constitute an interactional unit with any direct emotional meaning or impact. Keniston claims that, given the emotional constriction or frustration of the breadwinner at work, the family, especially so small a family, cannot satisfy his intensified or exaggerated emotional needs at home. Obviously, the wife faces analogous emotional frustrations, and then husband's and wife's intensified emotional needs become interactive.

There are no ready-made solutions for these emotional binds. Ideally, husband and wife, without abandoning their obligations to work and children, will move out into the community, into such things as church and civic activities, thus broadening the bases of emotional fulfillment. However, other less responsible solutions tend to destroy the equilibrium of the family: tension and fighting in the home, extramarital adventures, emotional constriction and insulation, and the vicarious emotional living mentioned above, made easy, for instance, by the proliferation of engaging but undemanding sports events on television.

Man's struggle for freedom has been the theme of much of his literature from the very beginning of recorded history. While this freedom is conceived of in more or less political terms, there has been a concomitant or parallel struggle for more interior forms of freedom such as emotional freedom. Today, even if men have been freed from the emotion-constricting slavery of Jansenism, Puritanism, and Victorianism (and certainly not all have), many have managed to shackle themselves with new bonds. While the prior slavery was enjoined in the name of morality and religion, the new slavery is imposed in the name of technocracy, progress, and production. Many have been duped into thinking that they are emotionally free, when all that has happened has been a change in the facade of their bondage. It is as if men were afraid to allow men to experience either themselves or their environment in an unfettered way and to institute communication with one another based on this experiencing. Rather, this is the unknown, the unknown is dangerous, the dangerous is to be feared, and the feared is to be resisted.

Some people are relieved when they are told that "feelings get in the way," for it justifies an already determined mode of interpersonal acting. Those who are guarded in their feelings toward others do not particularly want to become aware of these feelings. They would also prefer that others not feel strongly about them. It is thought uncivil, rude, unconventional, unwarranted, and even obscene to express feelings toward others. Emotional insulation parades under such euphemisms as "respect for others" and the "dignity of privacy." Sometimes the mentally ill are feared not because they express too little but because they express too much. Men who are afraid of feelings and emotions to begin with are utterly terrified when these are expressed without restraint. Perhaps the best symbol of man as emotional today is the polyethylene bag. Nothing gets in. Nothing gets out. He remains encased in interpersonal asepsis.

Men, by their inability or unwillingness to communicate deeply with one another, seem to foist upon themselves a state analogous to social deprivation. It would seem that such self-inflicted deprivation might well have effects analogous to those observed in studies indicating that variables such as monotony of environment and absence of sources of emotional gratification can cause intellectual inertia, impaired memory and concentration, insomnia, headaches, low-grade depression, and greatly increased appetite (with resultant weight gain). An interesting hypothesis is that similar symptoms found among relatively normal populations of our society reflect self-imposed estrangement from others, although such a macro-hypothesis would be difficult to verify. The contract group is a laboratory in which the participants come together to determine whether or not they themselves are victims of any form of self-inflicted social deprivation, and, if this is the case, to find ways of remedying such a situation by in-community activity.

It would be a simple but rather useless task to go on cataloging the evidence of man's sins against the emotional dimensions of his humanity. However clinical and anecdotal such evidence is, it is still compelling; but one need not prove what is self-evident. The popularity of organizations such as the National Training Laboratories, the Center for the Study of the Person, and of sensitivity laboratories in general dramatizes the plight of a people seeking deliverance from emotional bondage. Whether this deliverance takes place in responsible ways depends, at least partly, on the willingness of behavioral scientists to channel some of their talents into creative thinking and experimentation in this area. The growing laboratory-learning culture in our society offers any number of possibilities for such work. Perhaps a new era is dawning in the field of mental health, an era in which concern with prevention of illness is absorbed into larger concerns, such as self-actualization and community potential for human growth.

The Phases of Emotional Experiencing

The feeling and expressing of emotion, dealt with in a way that is meaningful to participants in an encounter group, has two phases: (1) awareness and impact; (2) reaction and expression. It is possible to short-circuit the experiencing of emotion at either of these phases.

Awareness and Impact

Emotion is preceded by a cognitive element, for the individual first *evaluates* the situation in which he finds himself. For instance, if he evaluates some aspect of his environment as hostile (such as a large dog running toward him), fear arises. Thus, the quality of a person's emotional life depends on the quality of his awareness. If, by either nature or nurture, a person's ability to be aware of what is happening around him is constricted, then his emotional life will also be constricted. On the other hand, if a person wants to grow emotionally, he must improve his awareness of himself and of his environment, especially his interpersonal environment.

The refusal or inability to be aware of self and others is characteristic of the emotionally disturbed. Usually the more disturbed a person is, the greater his tendency to cut off the emotional process at an earlier stage. The most severely disturbed, then, would sabotage emotional experience at the level of awareness. I have been in group-therapy experiences in which the most severely disturbed have been challenged merely to repeat the gist of what another had revealed in an emotionally charged disclosure. An "I-didn't-hear-what-he-said" answer typified the poverty of awareness that characterized these patients. This inner numbness or preoccupation with self effectively fended off any kind of affective contact with others. In a parallel way, the lack of awareness of self and others characterizes the more severe forms of the psychopathology of the average.

Furthermore, even if a person does not short-circuit emotional experiencing at its very roots—that is, at the level of emotion-producing stimuli, the level of awareness—he can still do so at the level of *impact*. It is true that at times feelings and emotions take us by storm; willy-nilly, we are flooded with fear, anger, sexual desire, or some other emotion. But even though a person remains more or less open to emotional stimuli at the level of awareness, he can still *learn* how to cut off the affective impact of that of which he is aware. While he realizes, in a rather detached, intellectual way, that he is encountering an emotionally evocative stimulus, and can even correctly identify the emotion or emotions that the stimulus is geared to evoke, he has learned how to neutralize the affect-evoking dimension of the stimulus. He has learned not to allow himself to react at all, or to react in such an attenuated way that there is no proportion between the stimulus and the strength (or weakness) of the reaction.

The neurotic has been described as a person who cannot be intimate even with himself; he is unable or refuses to let himself feel how he actually feels about himself and others. Studies have shown that the overly anxious tend to avoid sensation-seeking. They restrict their interest in the sensual, in excitement, in the new, in the strange, in the unpredictable. The same can be said of those who suffer from the psychopathology of the average.

Most of us are readily aware of the impact of the "heavy" emotions such as anger, fear, sexual desire, and depression, especially when they are strong and make us aware of their existence; but we are not as aware of or as open to their more attenuated or subtle forms, or to more subtly shaded emotions such as wonder, surprise, curiosity, ennui, and caution, to name just a few. The English language is filled with words referring directly or indirectly to these more subtle forms of feeling and emotion, but, though this might give some witness to the reality of such emotions, the reality does not seem to play the role it should in our interpersonal lives. For instance, when one person meets another for the first few times, he might "like" the other, but "like" merely summarizes a whole group of emotion-laden variables. Part of this liking is a rather wholesome and pleasing curiosity. He is attracted to the other and finds a certain delight in psychologically exploring the other or in engaging in mutual exploration. Again, the group is a laboratory for learning how to become aware of, experience, and enjoy the whole range of these more subtle emotions. Therefore, the laboratory must provide an atmosphere in which such subtle emotions are viable. Perhaps too often, sensitivity laboratories are rather heavy-footed, providing opportunities only for the more dramatic emotions.

Examples of Emotional Blunting in the Encounter Group. Some people in the group, because they are uncomfortable experiencing emotions, will tend to turn off the emotional process at the awareness and impact stage.

Example 1. There is the person who does not notice emotions in others, for he is uncomfortable with emotion wherever it is found, even when it is not directed specifically at him.

> *John*: Mary has really been uptight this last half hour, Bill; it has really been painful for her to deal with her feelings of inadequacy in this group. But Bill, you've been one of the very few who have said nothing to her.
>
> *Bill*: Gee, I must have missed it. I didn't see the pain. I guess I'm kind of out of it today.

When emotions arise in the group, people like Bill tend to drop out of the interaction. They "tune out" and do not allow themselves to be touched by the emotions of others. The emotions of others tend to arouse their own emotions, and they are not comfortable with their own emotions. These are the people with selective hearing, selective attention.

> *John*: Did you have any reaction, Bill, when Mary said she cared for you?
>
> *Bill*: Did Mary say that? I don't remember her saying that.
>
> *John*: She said it right before she started crying.

Bill: I knew you were feeling bad, Mary, but I didn't know you were crying. Sitting next to you like this, I can't see you head on.

This is a far cry from direct, concrete engagement with the other-as-subject-of-emotion:

John: Mary, I really notice a change in you. You are usually outgoing or at least quite ready to be contacted. But you seem depressed now, you've pulled away. I feel that you are distant from me right now, and it's making me uncomfortable.

Example 2. There is the person who almost always denies that any kind of emotion is welling up inside. He need not be lying, for he has learned not to react.

Eve: Kevin, you must really be frustrated. People have been confronting you about your lack of participation and you can't really see what they mean.

Kevin: Oh, no. I feel fine. It's just a question of defining what we mean by participation.

Perhaps Kevin should feel frustrated, for he is in the kind of situation that would produce some degree of frustration in most people, but he has schooled himself to feel practically nothing.

Sue: Peter, do you have any feelings about me?

Peter: I don't know. I'd have to sort it out.

Again, this kind of response is a far cry from direct emotional response to direct emotional contact:

Fred: Phil, I'm really angry with you because of the way you ignore me.

Phil: I ignore you because I see you as too strong and confronting. I'm afraid to be confronted by you. And I'm afraid of your anger right now. I don't know what I look like on the outside, but inside I'm shaking.

Example 3. There is the person who distorts the emotion of the other; he changes it into something he feels he can handle.

For instance, George is angry with Jane and manifests his anger in a variety of ways. Jane, unable to cope with George's anger, makes it out to be something else:

> *George*: I don't feel it right now, but I was really angry with you last meeting.
>
> *Jane*: I thought you weren't feeling well. You were angry?

Such a person can't identify his own emotions either. For instance, Sylvia somatizes her anxiety but refuses to recognize it as anxiety, for this is too disquieting:

> *Claude*: I know your colon gets into an uproar when you're upset, and it hurts me to see you suffer so. I'm thinking that you should deal more openly with the things that upset you.
>
> *Sylvia*: Oh, this happens whenever I hurry too much or don't eat right. I gobbled down my lunch today. It's not that bad; I ignore it.

Denial is an inefficient, if not dangerous, way of handling the emotional realities of life. Repressed emotion has a way of creeping out in destructive or at least unproductive interpersonal transactions or of creeping into the GI tract and causing physical problems.

Whatever the reasons underlying the tendency to eliminate or minimize the emotional impact of reality (people-reality), and whatever idiosyncratic patterns or syndromes this process might take, the participants in the encounter group commit themselves to determining the extent to which they are victims of such a process. To put things more positively, the group is a laboratory in which the participants try to allow affective reality to have as full and constructive an impact as possible.

Emotional Reaction and Expression

There are those who, although they are quite aware of the emotion-provoking stimuli in their environment and even allow these stimuli to have their impact, still short-circuit the experiencing of emotion at the level of expression. They have learned, however unconsciously, to control the emission of both verbal and nonverbal cues that would give others insight into their interior lives. They seem unemotional, but this does not mean that they are really devoid of emotion—frequently they are filled with emotion—but rather that they have learned to control its expression. Actually they are overcontrolled. All of us, at one time or another, engage in such control. Overcontrol is problematic only when it is resorted to at times when open expression of emotion would be a more responsible way of reacting, when it is resorted to frequently, or when it becomes part of a person's life style. The encounter-group laboratory provides an opportunity for its participants to examine the positive and negative aspects of the emotional control they exercise from day to day.

Some men choose to be emotionally incongruent; they look upon natural, nonverbal emotional reactions as too dangerous, too self-revealing, too intrusive, or too disruptive. And so they opt for a rather drab, expressionless, "archaic-smile" emotional style (or rather lack of style), in appearance and gesture, as being both proper and safe. The problem is that such emotional asepsis is less than human.

Finally, some people—even though they are aware of emotional stimuli, and even though they allow these stimuli to have their impact, and react in the sense that they allow their emotions to appear in public through their facial expressions and gestures—still truncate emotional experiencing to a certain degree by failing to make *active* use of these nonverbal forms of affective communication. It is the difference between merely allowing oneself to react and being involved in one's reactions, even taking delight in them, to the extent that they become part of one's active communication style. One can become active, in communication, even in such involuntary reactions as blushing, if, sensing one's reaction, one puts oneself "in" one's reaction in such a way as to say, nonverbally, by such actions as smiling, eye contact, a shrug of the shoulders or other gestures: "You have caught me, you have hit upon a vulnerable area, a point of shame."

Examples of Emotional Reaction and Expression.   Some of us, even when we are filled with emotion, find ways of keeping our emotions to ourselves, even though emotional expression might be quite appropriate.

Example 1. The emotionally guarded person may emit all sorts of cues that he is in emotional turmoil, but he will not translate his emotions into words.

> *Sarah:* I can see the tension, Joe, in the way you are hanging onto the chair. Your whole body is rigid. I'd really like to share what's going on inside.
>
> *Joe:* I really don't know.

In this case "I don't know" does not mean "I'm really not experiencing anything." It means rather "I'm not willing to say."

Example 2. Some of us think we are expressing our emotions openly, but we do it so subtly or indirectly that the emotional impact of our message is lost. Conversations like the following are very common in groups:

> *Jerry:* Mike, I thought you knew that I'm for you. That's the way I've felt in here from the beginning.
>
> *Mike:* Jerry, how was I supposed to know? Maybe I'm dumb, but I've really missed the signs. You really have not contacted me very much.
>
> *Jerry:* Well, I just thought you knew.

Support in groups is quite ineffectual unless it is expressed, and it must be expressed concretely and directly. In the example above, the cues for Jerry's support were so subtle or indirect that Mike did not pick them up.

Example 3. Some of us hide our emotions by giving misleading cues—that is, cues that either mask the emotion or, worse, indicate an opposite emotion. For instance:

> *Tony:* Mark, I think you've been very controlling in this group and I've resented it.
>
> *Mark:* This is the first time you've said that. Why didn't you tell me earlier?
>
> *Tony:* I did. For instance, I dropped out of the interactions when you controlled them.
>
> *Mark:* I'm disappointed that you didn't tell me about your resentment directly. Can we deal with it now? I'd like to.

It is unfair to think that others can accurately interpret ambiguous cues. It is infinitely better to say what you mean and what
you feel as directly and as soon as possible and appropriate.
Incongruent cues are even more difficult to handle. In one group,
one of the members was receiving some strong confrontation
about her failure to take initiative in the group, to move out of
herself and contact others actively and directly. She kept smiling.
Then one of the participants noticed a tear running down her
cheek. But she was still smiling! Many of the members then began
to feel quite guilty, for they had made her cry. But she did have
the responsibility of both contacting other members (the group
had this kind of contract) and of letting others know what kind
of emotion was welling up inside.

Some of us are not afraid of our emotions, so we don't have
to deny, mask, or distort them. We can use them directly in our
interpersonal transactions. We can feel the more subtle emotions
and both experience and communicate conflicting emotions when
they arise.

> *Gene*: I'm slightly suspicious of you, Nancy. I'm not sure you're
> willing to let me know fully how you feel about me. Since I
> don't know what you're about, I tend to hold back when dealing
> with you. I'm a bit unsure of myself with women, and this
> insecurity really pops up when I deal with you.

> *Carl*: I have feelings of trust *and* mistrust for you, Gloria, how-
> ever strange that sounds. I trust you in your nonverbal behavior.
> You don't try to hide your feelings—your face is very expressive,
> your whole body tells me whether you are bored or interested.
> When your face is serious I know you are serious, and genuine.
> But when you talk, your words, your verbal behavior, don't seem
> as genuine. I trust the nonverbal you, but I have a few misgivings
> about the verbal you.

Some of us have the skill to express our emotions and deal with
the emotions of others through nonverbal gestures. It is easy for
some to offer a hand to someone who is disconsolate or to embrace
someone who is crying. Others of us are extremely awkward in
expressing emotion in gestures. Since one of the functions of the
laboratory is to experiment with "new" behavior—that is, to try
to develop communication skills we don't have or have only in

a rudimentary way—it is appropriate to take some risks in trying modes of expression that are foreign to our style.

Hostility in Encounter Groups

One of the reasons that hostility is so prevalent, especially in the early stages of encounter groups, is that it is a natural reaction to the aimlessness of most groups. Some researchers find four "natural" stages in groups: milling around, hostility, identity, and work. However, if a group does not have effective goals and the structures to achieve these goals, of course the members will mill around. They are trying to find out what they should be doing. After a certain amount of milling around, the participants become frustrated and hostile. Some of the stronger members attack the leader (for he seems to be doing nothing to help the group out of its plight), while others begin to attack one another. Then the group begins to elaborate goals and finally gets down to the work of pursuing the goals that have been elaborated.

However, if members know what they are about from the beginning, then they do not mill about and become frustrated and hostile. Hostility, at least for some, is a relatively inexpensive emotion, but there is no research showing that it is a particularly constructive emotion. Hostility is used too recklessly in many encounter groups. A member may feel a certain degree of invigoration in venting his hostility on another (or on the group as a whole), but such expression does little to establish relationships of some closeness. Of course, if a member is legitimately angry, he should work his anger out with the person who is the object of his anger, but this is a different process from that of "dumping" hostility on others.

Hostility frequently expresses more than raw "againstness." Especially in group interaction, it can mean many things. (1) It may be a way of expressing one's individuality or showing strength in the group. This use of hostility, however, is relatively immature and usually characterizes only the earlier sessions of the life of the group. Real strength and individuality can be displayed in contractual ways. (2) For the person who feels threatened by the interaction of the group, hostility may be a defensive maneuver rather than a form of attack. (3) Planned hostility may be used as a dynamite technique to stimulate action during a boring session. (4) Hostility can also have a more subtle and constructive meaning: it may be an attempt to achieve some kind of interpersonal contact or intimacy. A number of authors have suggested (and some have conducted research that supports the hypothesis) that identification tends to follow aggression. It does seem to be a fact that sometimes after two people storm at each other, they tend to draw closer together. Perhaps the direct route to intimacy is too difficult, and the turmoil of the indirect route is all that is available.

To the degree that hostility is merely punitive it serves no growthful purpose in the group. Confrontation is meaningful in most cases only if it takes place in the context of care, concern, and involvement. The group that thinks it is going somewhere because it engages in a good deal of hostile confrontation is deluded. More will be said about this in the next two chapters, dealing with support and confrontation.

Emotion-Evoking Exercises

In some encounter groups, not only are emotions allowed to arise naturally as the participants try to establish and develop relationships of some closeness with one another, but occasionally exercises are introduced to further stimulate the arousal and expression of emotions. For instance, early in the life of the group the participants may be invited to mill around the room with their eyes closed. As they bump into people, they are to greet one another without using words. For most, this is a strange new experience that focuses on dimensions of the "greeting" process not usually considered. Exercises may be verbal or nonverbal, and they may or may not involve physical contact. Exercises involving some form of touch are usually powerful emotional stimulants, perhaps because many of us tend to avoid touch as a mode of communication.

Exercises involving touch are frequently quite diagnostic; they reveal areas of emotional constriction. For example, during a processing session that took place immediately following a few simple nonverbal exercises involving physical contact, one of the participants, who had been obviously rigid during the exercises, said:

> I really feel disturbed, not by the exercises but by my reaction to them. Over the past two years I have become much more at home with myself and with others. I felt that I was quite in possession of myself both on the personal and the interpersonal level. These exercises disturbed me because I experienced residues of rigidity in myself that I thought did not exist. I saw dramatically that not everything has been worked out, that I am still afraid of intimacy both with myself and with others.

Insofar as these exercises are diagnostic, they aid the process of communication in the sense that they put the participant in more effective (though sometimes painful) contact with himself. Such communication with self serves as a basis for more effective contact with others.

An interesting exercise for the person quoted in the example above (though it was not proposed at the time) would have been an exercise in self-dialogue. He would be asked to talk to himself, to engage in dialogue with himself, about the unpleasant discovery he had made. Perhaps the dialogue would have gone something like this:

> You feel pretty stupid, don't you, having to admit in front of all these people that you don't have your emotional life all together?
>
> I feel pretty stupid having to admit it to myself.
>
> What's going on, then, in your emotions? What's this rigidity all about?
>
> I grew up in a family that was rather constricted emotionally. I went to a seminary where I learned to be afraid of intimate contact, at least with males.
>
> But you said that you were unlearning all of that.
>
> I think now that I have been unlearning it in the head. My attitudes about intimacy have changed, but I still don't have much practice with real intimacy. I still have not gotten very close to anyone, especially to women.

You need practice, then.

"Practice" is an odd term. I need experience. I mean I have to contact people whom I sincerely care about and get involved with them. "Practice" isn't the word. What I do has to be genuine. I don't want to "practice" on my friends.

Well, then, do something. Start here in the group. And be genuine.

In self-dialogue, the participant can confront himself first before turning to the rest of the group for help.

Physical contact in our society is anxiety-arousing. First of all, it is an expression of intimacy, and many of us are afraid of intimacy. Second, physical contact in our society seems to be overidentified with sexuality; it is not seen as a universal mode of contact and communication. Anxiety runs very high when the exercise is so structured that the dyads are of the same sex, especially if both are male. "I would feel a lot better if my partner were a girl" says a number of things: "I live in a culture in which physical contact of male with male is more or less taboo"; "physical contact is, of its very nature, sexual"; "I consider intimacy as something intersexual." However, the laboratory is a cultural island; that is, it attempts to develop its own human culture apart from the cultural rigidities that exist outside the laboratory. In the case of exercises involving physical contact, the laboratory culture says this: "Physical contact is another channel of human communication. It can be so restricted as to communicate only certain dimensions of interpersonal living such as hostility (in acts such as shoving or striking) and sexuality (in *any* physical act showing interest or concern). Here we experiment with physical contact as a channel of communication. Our purpose is to see how many different human realities we can express through physical contact or through a combination of physical contact, nonverbal behavior that does not involve physical contact, and verbal behavior. Our purpose here is to grow interpersonally by involving ourselves with one another. Physical contact is one of the modes of human involvement. Here in the laboratory there is cultural permission to deal with it more freely than we could in day-to-day living. It is hoped that we can take advantage of this permission."

Physical contact can arouse strong emotion, especially strong anxiety. In my own experience, I have found that the way in which exercises are introduced to the group is of paramount importance. For example, in the first residential laboratory I attended, we had generally avoided exercises, especially nonverbal exercises involving physical contact, during the first half of the laboratory. Around the midpoint, we were involved in a session in which communication had noticeably bogged down. It seemed that we just could not get in contact with one another. We were sitting outside on a patio, when suddenly the trainer said: "I'd like to do something. Let's go inside." The anxiety level in the group shot skyward. My imagination ran wild: we were going inside because we were going to do something that should not be seen by others. Once inside, we sat around for a while, saying nothing. The trainer remained in a very serious, brooding mood. Our anxiety continued to mount. Finally he said: "I would like a volunteer." Again my imagination ran riot: volunteers are called for (especially under the "battle" conditions under which we were operating) only when the mission is dangerous. We remained silent and frozen in our seats. The trainer made another plea for a volunteer. Finally, with obvious trepidation, one of the men in the group (certainly not me) said that he was tentatively willing to try to cooperate. The trainer said: "Hold out your palms, I'd like to feel your strength and let you feel mine." They pushed against each other for a while, but the volunteer's willingness faded and he withdrew from the exercise.

What happened? The exercise proposed by the facilitator was really a simple, fairly nonthreatening one, but the way he introduced it created such anxiety that it practically immobilized the group. First of all, there was reluctance to go inside to do something that could not be done outside. As we moved inside people stopped talking to one another. Once we got inside, we sat around looking at the floor, avoiding both verbal and nonverbal contact with one another. I felt a great deal of anxiety because I thought that we were going to engage in some kind of nonverbal exercise. This was my first laboratory experience and up to that point in the laboratory we had no physical contact with one another. The solemnity of the facilitator, the hush that had fallen over the group, the change in location, the fear of the unknown— all of these raised my level of anxiety greatly. What should we

have done? We should have gotten in touch with our panic.
I, for example, should have said:

> I'm extremely anxious right now. I don't know why we came
> in here, but I think it was to engage in some of the exercises
> that I hear are taking place in some of the other groups. I feel
> that a lot of people in this room are afraid; at least I am and
> I'd like to check it out with the others. We're not even talking to
> one another. What's going on?

Perhaps the facilitator should have said something like:

> We're really mired down this afternoon. We haven't had any exer-
> cises in this group so far. I don't think we needed them. Let's
> go inside where we will be a little more comfortable—we won't
> have to be thinking about others staring at us—and see if we
> can come up with an exercise or two that will help us break
> up the communication blocks we are experiencing right now.
> We'll start with volunteers, so don't think that you're going to
> be pushed into doing something you don't want to do.

Neophytes are going to be anxious. A certain degree of anxiety
is helpful; it can keep people at the task of involving themselves
with one another. But excessively high anxiety usually immobi-
lizes.

Obviously exercises are not meant to be a way of life. They
partake of the general artificiality of the laboratory setting.
However, if they make a participant pause and reflect on some
dimensions of his interpersonal life, if they show him some of
his unused potentialities, if they enlarge his area of freedom with
himself and with others, then they serve their purpose well.

Some Cautions

I'm not sure that "caution" is the right word, for it might
merely perpetuate myths about the danger of emotional expression.

1. Creative expression of emotion in interpersonal situations
is a skill. As with any skill, some kind of "practice" is needed;
it will not develop merely by "thinking on't." Awkwardness is
to be expected in the beginning.

2. Emotional expression is the kind of skill that is developed best in an atmosphere of acceptance, understanding, warmth, and support. If the group fails to create a climate of trust and support, then its members will not take the risks necessary to develop skills in emotional expression.

3. Skill in expressing emotion is not worth a tinker's dam unless the emotions expressed are genuine. The purpose of the group is to produce not actors but human beings.

Man's Translation of Himself into Language

The encounter group places emphasis on effective interpersonal communication through human language. I stated above that a great many people suffer from emasculation in their emotional life. Now I'm asserting that many people also suffer a concomitant emasculation in the quality of their verbal communication, in their ability to use language as a mode of interpersonal contact. Language reflects the psychopathology of the average. Many normal men fear the communication process because of more or less normal fears of involving themselves deeply with others. They neither pour themselves into their language in interpersonal situations nor expect others to do so. Language must remain on a safe level. They habitually put filters between what they really think and feel and what they say. "If I keep my deepest thoughts to myself, the best in me and the worst in me, if I prattle on with others on a safe level, then I will never risk being ridiculed." Most of us don't say this out loud or even to ourselves.

Language is an instrument by which man examines the world about him. If he is afraid of this world, then his language will be anemic and feeble, but if he loves the world and allows himself to be challenged by it, his language will be strong and searching. To adapt a phrase from Wittgenstein, the limits of a person's language are the limits of his world.

Brian Friel's entire play *Philadelphia Here I Come* is based on the distinction between what the leading character really thinks, feels, and would like to say and what he actually says. In the play, there are two levels of conversation—the vague, hesitant, compliant, failed bravado of the son about to leave his father in Ireland to seek a new way of life in the United States, and the vigorous speech of the son's "inner core" (played by a separate character). The pity of it all is that, although the audience is electrified by what the "inner man" says, it knows that his speech really dies (and in a sense the son dies with it) because it is never spoken. The man who chains his language chains himself.

The encounter group is a laboratory in which the participants can experiment with the potential of language. The purpose of what is said here is not to glorify language, for language is sometimes a sensitive instrument but other times a clumsy tool of communication. But when a person enlarges the possibilities of his language, he enlarges his own possibilities. The laboratory gives him the opportunity of extending the range of his language in order to contact himself and others at deeper levels. In the safety of the laboratory, he can run the risk of pouring himself into his language in ways in which he does not in everyday life.

Different Kinds of Language

Language that Brings About Human Contact. This kind of language permits a person to translate his real self into words and by doing so to make deeper contact with his fellow man. It is language filled with the person who is speaking. Negatively, it implies a refusal to use speech merely to fill interactional space and time or as a smoke screen behind which to hide.

The ability to speak in such a way as to make human contact must be clearly differentiated from the ability to speak fluently or elegantly, for both fluency and elegance are at times used to camouflage rather than reveal one's identity. If the encounter-group member is to develop new ways of being present to the other members of the group, then he must discover "new" ways of speaking.

Commercial Speech. "Commercial speech" refers to the language of the marketplace, the use of language in the commercial transactions of men. Such language is lean, utilitarian, pragmatic; it deals with objects rather than persons for it is a medium of exchange rather than of interpersonal contact. Much of such language today is left to computers. It would be of no interest to us here were it not for the fact that there are people who use commercial speech as their principal mode of speech in interpersonal transactions. They see people as objects to be manipulated, rather than persons to be contacted, and this is reflected in the quality of their speech.

> *Luke*: I like groups.
>
> *Martha*: What do you like about them, Luke?
>
> *Luke*: It is fascinating to see all the things you read in social psychology texts about group interaction come to life right around you. For instance, when John and Mary went after one another this morning—you expect *just that* at this stage of the group!
>
> *Martha*: You don't sound too involved with the individual people.
>
> *Luke*: There are many different ways of getting involved.

Commercial talk is too distancing and antiseptic for the encounter group.

Cliché Talk. "Cliché talk" refers to anemic language, talk for the sake of talk, conversation without depth, language that neither makes contact with the other nor reveals the identity of the speaker (except negatively, in the sense that he is revealed as one who does not want to make contact or does not want to be known). Cliché talk fosters ritualistic, rather than fully human, contact ("Do you think that it is really going to rain?"—"The way they're playing, they'll be in first place by the first of September!"). Cliché talk fills interactional space and time without adding meaning, for it is superficial and comes without reflection. Perhaps it is the person who is overcommitted to maintenance functions (see Chapter 1), a person who is either unaware (because he lacks the requisite social intelligence) or afraid of possibilities for further interpersonal growth, whose speech will be predominantly cliché talk.

Consider the differences between the following two conversations dealing with the weather.

> *Warren*: It's really chilly today, damp—real winter.

Steve: It's very humid, a wet kind of cold. They say that this is the "wettest" winter on the books.

Warren: All sorts of records are being broken this year. It has been both too hot and too cold for me.

The other possibility is this:

Cynthia: It's really chilly today. It's getting right into my bones and marrow.

Boris: What does a day like this do to you? I get terribly bored sometimes looking at all the gray. At other times, like today, something in me responds to grayness and lifeless trees. I feel my finiteness today. It half makes me miserable and half makes me wonder what life is about.

Cynthia: That strikes something in me! The dampness is like death creeping into my bones and marrow. I shudder and yet it is not entirely unpleasant. I've been thinking a lot about death lately, yet it isn't morbid.

Boris: You and I are a lot alike in some things—somehow we find life in grayness and death. I feel close to you right now, but at other times I feel that we are miles apart.

Cliché talk is just words, while language-that-contacts is speech filled with the speaker. Some people speak endlessly about themselves and say nothing (if they were really disclosing themselves, others would not find it boring). They say nothing about themselves because they have no real feeling for themselves—they are deficient in the emotional dimension of life—and could hardly be expected to relate what they don't experience.

Speech as a Weapon.   Sometimes language is actually used to destroy interpersonal contact rather than foster it. There are a number of forms of speech that are really violations, rather than uses, of language. For instance, in the heat of anger, language can become a weapon. When married people stand shouting at each other, their language has more in common with the sledge-hammer than with a process of communication.

Translating Nonverbal Messages

When two or more people are talking, there are usually at

least two levels of communication: (1) what is conveyed by the verbal interchange (the content mentioned above) and (2) a variety of other messages that are transmitted in a number of different ways—for example, the qualities of the verbal exchange itself such as speed, tone, inflection, intensity, and emotional color, and nonverbal cues such as eye contact, bodily stance, facial expressions, and gestures. These messages are "metacommunicative processes," the purpose of which is to interpret or classify the content of the verbal message or to send a parallel message more or less unrelated to verbal content. These metacommunications can even negate or deny the explicit meaning of the verbal message (for instance, it is a rather common occurrence to hear "no" on an explicit verbal level and at the same time to experience "yes" on a metacommunicative level, the latter being the real message).

Sometimes entire encounters take place in which the explicit verbal level is relatively meaningless, for it is only a sterile carrier for an exchange of metacommunicative messages. Language becomes a pastime, something to do while the reality of the encounter takes place at a different level. It would be an interesting experiment to have groups of two or more talk to one another for a half hour and then have each individual write down three or four of the more important nonverbalized messages that he thought, on reflection, he had communicated to the others and three or four of the messages that he, again on reflection, thought that each of the others in the group had sent out on a nonverbal or metacommunicative level. The hypothesis would be that there would always be some correspondence between messages believed sent and messages believed received (that is, people would "read" one another, if asked to do so, on a metacommunicative level, even though they might seldom verbalize the fact that they have either sent or received such messages). Still, people would differ in their sensitivity to metacommunicative messages and in the number and kinds of messages sent out over nonverbal rather than verbal routes.

Language and Emotion

In human affairs there seem to be two highly prevalent, though probably not growthful, ways of handling strong feeling—both positive and negative feeling. Actually, both are ways of avoiding, rather than handling, emotion in transactional situations.

The Suppression of Feeling.   The safest way of handling strong feeling is to suppress it. Perhaps "conceal" is a more accurate word than "suppress," for hidden emotion does make itself felt under a number of disguises. For instance, if a person suppresses or conceals his anger, it frequently comes out in a number of deceitful ways, such as coolness, unavailability, snide remarks, obstructionism, and other subtle forms of revenge. Feeling has not really been suppressed; rather, it has been translated into a number of nongrowthful activities that are difficult to deal with precisely because of their underground character.

One study describes a work camp in which, because of the philosophy and religious convictions of the members, the prevailing atmosphere was one of friendly and gentle interactions. Since the members disapproved of all kinds of aggression, both physical and verbal, a problem arose with respect to the handling of the minor antagonisms that arose daily and tended to interfere with the work to be done. Meetings were held, but problems were discussed in a most abstract and intellectualized way. Because of the failure to institute real emotional communication, the antagonisms persisted, much to the dissatisfaction of everyone. But an intellectual approach to a nonintellective situation was bound to fail.

Language Unafraid of Emotion.   Let us suppose that once George has been angered by John, he says something like this: "John, I am really angry with you. I could try to swallow my anger or I could blow up, but I don't think that either of these would solve anything, because I think that in a way my anger is really *our* problem, yours and mine, and I'd like to talk it out with you. How about it?." Such a tack (especially if the stylized way in which it is presented here is overlooked for the moment) is rarely employed, for it demands too much honesty and one runs the risk either of refusal or of disquieting discoveries about oneself. It also demands dealing with feelings instead of relinquishing them in one way or another. George remains angry, but now his anger becomes a point of possible contact instead of just an abrasive force. Sometimes a person has to choose between the pain of talking out another's hostility toward him and the discomfort of being the victim of a dozen covert expressions of hostility so rationalized that it is impossible to get at them.

A Final Word. Laboratories are admittedly artificial, yet many meaningful learnings take place during them. It is true that the whole atmosphere forces the participants to deal with interpersonal emotional issues, but that is what the laboratory is designed to do. People can still be phony in the laboratory, they can play emotional games, manufacturing emotion because they think that it should be manufactured or refusing to give expression to emotions actually felt because they are so unfamiliar. Still, in my experience, a good deal of authentic emotional expression takes place in encounter groups—enough to dramatize to most participants some of their areas of strength and of deficit in the emotional dimensions of their lives.

When emotion finds expression in human language, both verbal and nonverbal, it is poetic in the deepest sense of that word. When meaning and feeling become artfully one in language, the result is poetry. In human dialogue, when words are meaningfully filled with human emotion, when feelings and emotions find creative expression in human language, the result is also poetry in a sense. It is strong language as opposed to the anemic language of too many of our encounters. Such strong language or poetry is not very welcome on the interpersonal scene, for it is an anti-manipulative and anti-game form of communication in a manipulation- and game-prone culture.

Our present culture is probably not ready for a sharp rise in direct, strong language. The character Jerry in Albee's *The Zoo Story* is somewhat disconcerting to the average reader, for people are not accustomed to dealing verbally with reality on the level that he deals with it. Jerry is resented because he feels too much and because he translates what he feels into language. He is too direct. Anyone, then, who is responsibly and intelligently poetic in his encounters must expect to experience a certain amount of rejection from those who cannot tolerate intimacy.

Chapter 7

Active Support

If the encounter group calls for such behavior as self-disclosure in a demanding sense ("story" rather than "history") and a willingness to deal openly with the feelings and emotions that arise in the give-and-take of group interaction, then there must be an atmosphere that not only sustains but also *actively* encourages such behavior. In a word, the encounter group demands a climate of *support*. The problem is that our attempts to give solid, non-cliché support are too often clumsy and ineffective. When faced with other than the superficial dimensions of others—when faced with their pain, their anxieties, their peak and nadir experiences, their strong emotions, their successes and their failures—we fumble around, babble inanities and social clichés ("I know how you feel"), or take refuge in silence. In a laboratory in interpersonal relations, one of our goals is to learn how to react humanly to the more dramatic or emotion-charged dimensions of others. It is an opportunity for us to face and deal with our deficiencies in both giving and receiving support.

The importance of support in the encounter group cannot be overstressed. If the group as a whole is truly supportive—and we will have to see just what this means—then there will also be a climate of psychological safety. The group must be a safe place for the individual to expose his feelings, lower his defenses, and try out new ways of interacting. But the group must be more than "safe." It must be a place where the individual participant is actually encouraged to engage in the kinds of behavior described in this book. Support, then, is much more than just a climate of permissiveness. It is a climate of active concern.

Listening as Empathic Support

For our purposes, listening means becoming aware of all the cues that the other emits, and this implies an openness to the totality of the communication of the other. In the encounter group, such openness requires being aware not only of individuals but also of the mood of the group as a whole. Perhaps listening to the group is even more difficult than listening to individuals because it demands an awareness of subtle interactional patterns. Ideally the facilitator is already sensitive to these patterns, and one of his functions is to point them out to the other members. Listening demands work, and the work involved is difficult enough so that the effort will not be readily expended unless the listener has a deep respect for the total communication process.

One does not listen with just his ears: he listens with his eyes and with his sense of touch, he listens by becoming aware of the feelings and emotions that arise within himself because of his contact with others (that is, his own emotional resonance is another "ear"), he listens with his mind, his heart, and his imagination. He listens to the words of others, but he also listens to the messages that are buried in the words or encoded in the cues that surround the words (the "metacommunications" of the other). He listens to the voice, the demeanor, the vocabulary, and the gestures of the other, to the context, the verbal messages, the linguistic patterns, and the bodily movements of the other. He listens to the sounds and to the silences. He listens not only to the message itself but also to the context, or in Gestalt terms, he listens to both the figure and the ground and to the way these two interact.

Some examples will help clarify this kind of active listening.

Charles: Your words say you have no problem with me, Clara, but when you talk to me there is very often an edge in your voice.

Clara: Well, some things about you do bother me, but they seem so small that I didn't want to bring them up.

Kurt: The way you are sitting back stretched out on the chair bothers me, Dan. You say that you're here, but your body tells me that you are only partially here.

Dan: Maybe I *am* only partially here. Your interaction with Sandra has been pretty long, pretty exclusive, and therefore, for me, pretty boring.

Kathy: You haven't said anything for a long while, Burt, yet I don't feel you are someplace else. I have the feeling you're very uncomfortable with what's happening here.

Burt: You and Kathy and John have been talking to one another with such care and concern. I just didn't know how to get into the conversation. I'm not used to this. My conversation is usually so superficial if not actually cynical. I admire what is happening and yet I feel miserable because I can't seem to be a part of it.

Christopher: One senses an empathic understanding here that is crucial to the kinds of interaction that characterize these groups.

Bridget: Your words sound like a lecture, Chris, but I take them to mean that you're glad that I get inside you somehow and share your world. I'm glad that you share mine. You do, you know.

This kind of active listening is the basis for empathic understanding.

The Nonselective Character of Total Listening.   Total listening is, in a sense, nonselective: it encompasses all the cues emitted by the other, even those that the other would rather conceal and those the listener would rather not hear. For instance, the weight of an obviously overweight person is a cue to be reckoned with, for through it the other is saying something to those with whom he interacts. The message may be "I am frustrated" or "I don't care about others" or merely "I have poor self-control," but it is a message that should not be overlooked. In a group therapy session in which I was an observer, the therapist asked the wife of one of the inpatients (the patient himself refused to attend the sessions) what she thought she was trying to tell others by her obvious overweight. The therapist listened to a cue, confronted the woman in a firm, kindly, responsible way, and succeeded in instituting a dialogue that proved quite useful. Therefore, good listening demands both subjectivity—that is, engagement with the other—and objectivity—disengagement from the other—in order to pick up both positively and negatively valenced cues. The good listener is sensitive to what is and not just to what should be or to how he would like things to be.

Active Listening.   It becomes quite apparent that the good listener is an active listener, one truly engaged in the communication process, one who goes out of himself in search of significant cues emitted by others. Listening, then, is facilitated if the listener is actively interested in others. The person who is an active listener is much less likely to stereotype others or to become guilty of univocal listening. Perhaps an analogy would make this a bit clearer. Everytime Brahms' Second Symphony (a favorite of mine) is played, the untrained ear hears only the Second Symphony; it is quite a univocal experience. The individuality of different orchestras and different conductors and the nuances of different tempos and accents are all missed. However, while there is only one Second Symphony, it can be played with quite different—and distinguishable—nuances. Similarly, John Doe is only John Doe, but John Doe, too, has different nuances of orchestration at different times, and these nuances will be missed by the untrained, passive listener who finds it more comfortable to deal with him as a stereotype in univocal terms. The active, searching listener, who is open to all the nuances of John Doe, will more likely pick up many of these cues. This openness to nuance, however, does not imply that the good listener is skilled in analyzing the other, for analysis often means reducing the other to a whole series of stereotypes, and sometimes this mistake is worse than the first.

Some Obstacles to Effective Listening.   A number of things can get between the one communicating himself and the listener. Awareness of them may help the listening process. *Self-consciousness*: if a person is preoccupied with himself as an interactant, he will find it difficult to catch what you are saying.

The dreamer: such a person gets lost in his own reveries. He is thinking about what you said to him ten minutes ago and is lost to the interaction. The solution is obvious. He should share his thoughts with you as they come up instead of engaging in distancing cud-chewing. *Message anxiety*: the content of the speaker's message might be such as to arouse both the speaker's anxiety and the anxiety of those listening. This is often the case when the message is suffused with emotion. There is no simple solution to this problem. When you notice that the speaker's message is arousing his and/or your anxiety, be careful. Message distortion is just around the corner. *The long speech*: the listener tends to lose parts of longer speeches, especially the middle portion. Again the solution is obvious. Don't give long speeches or allow them to be given. The group is for dialogue, not speeches. *Reductive listening*: the listener tends to modify a new message so that it sounds like previous messages. This is quite common and quite unfair to the speaker, for it is a refusal to admit that he can change. People do change during the course of the group. If you cannot hear these changes, you can hardly accept or support them. *Hearing what one expects to hear*: this needs no comment. *You agree with me*: the listener tends to modify messages so that they are in better agreement with his own opinions and attitudes. If you are to be a good listener, you must learn how to relax your defenses a bit in order to be open and willing to explore the new. *Black-or-white listening*: we tend to evaluate messages as we hear them in terms such as "good, bad," "like, dislike," "approach, avoid," "beautiful, ugly," and the like. Such instant evaluations obviously cloud the communication process. One way of fighting this tendency is to admit to the speaker that your automatic evaluation process is getting in the way: "I didn't like what you first said and it got in the way of the rest of the message; it would help me to check your message and my feelings with the group."

It is difficult to be an unbiased listener. The pitfalls described here cannot be avoided entirely; they can only be minimized. All the blame for poor listening, however, cannot be laid at the door of the listener. Whenever a participant speaks but really says nothing, then the quality of listening in the group will go down. As a general rule, the speaker gets the kind of attention he deserves. Some participants speak vaguely and evasively because they do not want to be read by their listeners; they thrive on the haphazard listening they receive. Such tendencies should become the object of confrontation in the group.

The Empathic Listener: The One Who Responds. The good listener, then, is the active listener, the one who is as receptive as possible to all cues and messages generated in the group, the one who is aware of and committed to combatting the variety of obstacles within himself and others to effective listening. The empathic listener, however, goes further than this; he realizes that the proof of good listening lies in the way he responds to those to whom he is listening. Accurate empathy is absolutely essential to growthful interpersonal relations. It requires you to get inside the other, to view the world through his eyes, and to *communicate* to the other your understanding of him. Accurate empathy demands, at the minimum, the ability to let the speaker know that you understand the obvious meaning of his words. A higher level of empathic understanding involves the ability to go beyond the words, the ability to put together the cues found in tone, gesture, and context and to respond to the fuller message of the speaker—the message that goes beyond his words.

> *Chester*: I feel that we're not really doing much in this group. We're not really getting in contact with one another.
>
> *Don*: You do seem to be pretty frustrated with the way things are going, Chet. You seem especially disappointed in me. In fact, I feel you are more disappointed in me than in the group or the process of the group, because you've tried to contact me several times and I haven't been very responsive.
>
> *Chester*: Well, it is you. You bother me the most.

In this example Don has listened to all the cues that Chester has emitted and he responds, accurately, not just to Chester's words but to the context of the words. He understands what is going on inside Chester and he communicates this to him. His expressed understanding, since it is caring and accurate, allows Chester to become more concete about what is bothering him.

If you want others to involve themselves growthfully with you, you must let them know you are listening to them. The best evidence of your listening is the quality of your response to what others say both verbally and nonverbally.

Acceptance and Encouragement:
Creating a Climate for Interpersonal Growth

There is universal recognition that some degree of acceptance and warmth in interpersonal relationships is absolutely essential for psychosocial growth. This need begins at birth and, although it might undergo certain transformations throughout the maturational process, is still a strongly felt need during old age.

The need for a supportive climate in encounter groups is crucial. One of the potential drawbacks of groups is the tendency not to supply sufficient support, especially in early meetings, to enable members to cope with the stresses generated. The fear of nonsupport is one of the greatest anxieties experienced by group members. Participants foresee quite easily the potential pain of the laboratory experience, but what they do not or cannot foresee is a climate of support that will render painful interaction not only tolerable but even deeply stimulating and satisfying.

Two Phases of Supportive Encouragement. Supportive encouragement has two phases: *antecedent* (the "before" phase) and *consequent* (the "after" phase). Anyone who directly or indirectly encourages the participants to engage in the kinds of behavior that make encounter groups successful—that is, the kinds of behavior outlined in this book—is engaging in *antecedent* support (he supports the other in order that the other *will* engage in growthful behavior). For example, if the facilitator models good encounter-group behavior—let us say he engages in the kind of self-disclosure that aids the process of establishing and developing relationships—then his behavior is a form of *antecedent* support, for it encourages the other members to engage in the same kind of behavior. On the other hand, recognition given to participants for actually engaging in good group behavior is *consequent* support. It is saying, in some way, to the good participant: "You are an asset to this group." Since consequent support cannot be given until someone acts, let us first take a look at the varieties of antecedent support.

1. *"I accept you because you ARE": Acceptance as Antecedent Support.* If the encounter group is to work, there must be a climate of basic mutual acceptance. What is necessary is the kind of acceptance ideally owed another simply because he is a human being. It is a willingness to let the other be who he is and what he is, but it is an active, concerned letting the other be rather than a detached "not giving a damn" what the other is like or what he does. It means allowing the other to have the psychosocial lifespace that he needs in order to be himself as fully as possible. Negatively, it means a refusal to exercise various sorts of control over the other, a refusal to demand that his life style conform, generally or in specific aspects, to one's own—for example, with respect to style or modes of interpersonal interaction and to a value system. Acceptance implies an active allowing the other to be different from oneself, "active" here meaning that A's interaction with B should actually foster B's otherness, his differences, his unique way of being. In a laboratory-training group, it is as if each member of the group were to say to every other member: "You have a value that neither I nor we collectively either determine or can abrogate. We recognize you as *being* this value."

In the encounter group often a simple direct statement of supportive acceptance is very helpful.

> *Frank*: You've mentioned, Jack, that you see me as a rather strong person. I hope that doesn't make you afraid of approaching me. First of all, I have a number of weaknesses. But more important, I feel I'm on your side. I know it's more difficult for you to express yourself than it is for me, but I want you to contact me—and the others here.

> *Jack*: It gives me some courage to hear you say it.

Too often we assume that others know that we are "for" them. It should not be assumed in the group. Support should be expressed in a variety of ways.

At the very minimum, acceptance demands that one allow the other to express the ways in which he is different, different from the other members of the group and perhaps different from the images of man that are currently acceptable in our society. But if such acceptance is to be active, the members must be willing

to say to one another: "Your actions cannot be so different or
bizarre as to open an unbridgeable gap between you and me."
Insofar as acceptance is active, it approaches what Fromm (1956)
calls "brotherly love," the most fundamental kind of love. This
love means such things as care, respect, and the desire to further
the life of the other. Such acceptance or love tends to disregard
status, for it is a love between equals.

It would be foolish to assume that it is easy to develop a feeling
of "brotherly love" for everyone and to express it. Some people
make it hard for others to love them. The cold, the indifferent,
the hostile, the obnoxiously cynical cannot be approached in the
same way as the open, the trusting, the concerned. In the group,
express what you feel, but be open to change on the part of the
person who makes it difficult for others to get close. For instance:

> *Ron:* Your cynicism, Barb, turns me off. It really keeps me at
> a distance. I get so angry sometimes that I want to tell you to
> go to hell. I know there is more to you than your cynicism. I
> wish you could lay it aside and let me know you in a different
> way.

Care can only be offered. There is no assurance that it will always
be accepted.

I do not think that deep, human acceptance of others necessarily
implies or is synonymous with approval. In the example above,
Ron accepts Barb, but without approving her cynicism. If I am
sincerely interested in human growth, then I do not expect my
friends to accept me in the sense that they overlook, discount,
or even approve of modes of acting that are antithetical to their
values or that I myself am dissatisfied with. Wholesale approval
of another is often a way of expressing radical noninvolvement
or nonconcern. If I really don't care for you, then I can readily
approve whatever you do, because it costs me nothing to do so.
Or if my unlimited love for you is really an unlimited need for
your affection, then I might be ready to do whatever is necessary
to maintain my relationship with you. There is no evidence that
unlimited "love" or approval is growthful. Indeed, there is evidence
to the contrary.

2. *Antecedent Support as an Expressed Sense of Solidarity in
the Human Condition.* An acute awareness that man, for all his

splendid accomplishments, often not only chooses unwisely but doggedly adheres to self-destructive choices—this awareness, expressed by group members in such a way as to make it evident that no one is exempt from human folly, is a form of antecedent acceptance. The sooner a member gives sufficient cues to indicate that he is open to both the heights and the depths of the human condition, the sooner will he find himself in community in a spirit of mutual trust. The person who can say "We are a microcommunity of men participating in both the wisdom and folly of man" expresses a kind of solidarity with others in the human condition that is both a statement of acceptance of others and a plea to be accepted by others.

If mutual acceptance is to be accomplished, then the participants must lay aside not only formal status roles (such as psychiatrist, psychologist, clergyman, manager, teacher) but also any role that would interfere with free contact among group participants. For instance, if someone assumes a quasi-role based on the supposition that "problems put people in categories, problems divide," and if he translates this role into interaction in the group, saying implicitly or explicitly "I've listened, but my hang-ups are not yours nor are yours mine; we're playing this game in different ball parks," he is assuming a role that demands that he reject others and that others reject him. This kind of psychological or interactional distance, no matter how subtle or covert, inhibits mutual acceptance and, therefore, limits or interferes with the kind of trust that is absolutely essential in the group if all the participants are to contact one another freely. This distance works in two directions. The one who says, at least by implication, "My problems are not yours; they make me less than you, they set me apart from you" sets himself apart from others, making it very difficult for them to provide him with any kind of support. Though he does this in order to make himself less vulnerable to rejection, he defeats his purpose because he creates an atmosphere in which support is not viable. If he also adds a poor-me element, he complicates matters further, for he both refuses support and, at the same time, tries to extort it. This makes the rest of the participants ambivalent toward him, if not angry. On the other hand, if a participant takes the attitude toward another: "Your problems are not mine; your problems set you apart from me," then his "support," if he gives it at all, will smell of condescension, and the one being "supported" will resent being patronized.

Sometimes a simple statement about one's own vulnerabilities can have a very supportive effect in a group.

> *Tom:* All of us here seem so afraid to reveal ourselves. I know I'm afraid. I have a facade, but actually I'm a very timid man. I'm afraid to reveal myself because I think you are going to laugh at me. Not out loud, but inside. I'm fearful of a lot of things in life. I don't reach out and seize life as I should. I let life happen to me.

> *Helen:* I'm not a very accepting person and I don't want to pretend that I am. I have a whole history of turning people away from me. Frankly, I'm here to try to deal with it.

> *Ken:* Then do you want me to challenge you when you do things to turn me off, to get rid of me?

> *Helen:* Please.

When I reveal my vulnerabilities, I express my solidarity with others in the human condition, I make it easier for others to reveal themselves, and I invite others to confront me and challenge my unproductive forms of interpersonal behavior.

3. *Availability as Antecedent Support.* Friendship—and this includes the beginnings of friendship in human-relations groups —may be defined in terms of availability. Friends are mutually available, and the *degree* of their availability defines the strength or the depth of the relationship. Some distinction, however, must be made between physical and psychological availability. Physical availability refers to the spatiotemporal dimensions of the relationship. There is a high degree of physical availability if one person spends a good deal of time with another, if he remains geographically close, if physical presence of some sort (even contact by telephone) can be easily achieved, or if more intimate kinds of actual physical contact are a dimension of the encounter. But, as important as physical availability is for friendship, psychological availability is even more important. Physical and psychological availability are separable, and the latter is the more difficult to define. First of all, any kind of availability, whether physical or psychological, can be either active or passive. For example, A

might invite B to spend some time with him or, on the other hand, he might merely allow B to be with him. These would be examples of active and passive availability. A person is actively available in a psychological way if he takes the initiative in sharing himself—his deeper thoughts, concerns, feelings, and aspirations —with another. If he merely allows the other to share such things with him—that is, if he is a more or less willing listener to the confidences of another—then he is also psychologically available, but passively so.

It is evident that there are all sorts of combinations and degrees of availability, both active and passive, physical and psychological. For instance, a prostitute might be actively available on a physical level, but not psychologically available at all. Perhaps the ideal marriage, in terms of availability, is one in which the partners are utterly psychologically available to each other and in which mutual physical availability is worked out according to the individual needs and responsibilities. Marriage partners come together with a frequency and an intimacy of contact not available to others, and their physical intimacy symbolizes, promotes, and enriches their mutual psychological availability. Both their physical and psychological intimacy reveal how deeply they are "for" each other, how deeply each wants to support the very being of the other. In like manner, failures in marriage, and in friendship in general, can be conceptualized in terms of failures in physical and psychological availability, both active and passive.

In the encounter group, then, members are supportive to the degree that they become available to one another. At first glance, it would seem that they are always, as long as the group is in session, physically available to one another, but even in the group itself there are degrees of physical availability. There are certain physical cues—for example, looking at the other, modulations of voice—that indicate psychological availability or the beginnings of it, and these cues may or may not be present. One of the main purposes, in my opinion, for using nonverbal contact exercises in the group is to allow the participants an opportunity to use physical contact both to stimulate and symbolize their psychological availability to one another. Such exercises, however antecedently anxiety-arousing or silly they may seem to be, are actually serious and fear-reductive, for they usually reveal others as more psychologically available than one had realized. The sooner the participants become available to one another, and the more deft they become in finding ways in which to reveal or give evidence of this availability, the sooner will they create a climate of trust that will support more than superficial manifestations of the modalities of self-disclosure, expression of feeling, and confrontation.

4. *Participant Genuineness as Antecedent Support.* The behavior of the genuine person is antecedently supportive. He is the person who is himself, openly and fully. He is open to all types of experiences and feelings without being defensive or retreating into safe roles. He expresses this openness both verbally and nonverbally. Meeting a genuine person is a refreshing experience because we deal with the person himself and not a polite facade. A congruent person is one who can be instinctively trusted. Because he is open to his own experiencing and the experiencing of others, he deals with others fairly. We know where we stand with such a person because he is not afraid to tell us how he feels about us. He tells us the good, the bad, and the indifferent, but does so with care and concern. Such genuineness is obviously an ideal. The group that has a few members who are relatively genuine is fortunate, for they add an atmosphere of openness that is infectious, an atmosphere that encourages others to venture from behind their facades.

5. *Trust Formation as Antecedent Support.* Group members will differ in their ability to trust one another because they come with different developmental histories, different learning experiences involving trust, disappointment, and mistrust. It is important for the participants to realize this as they attempt to create a climate of trust. The participants' unresolved feelings of fear and distrust, even though these may be buried and denied, are formidable obstacles to the formation of an intimate community that fosters interpersonal growth. Acceptance and trust must precede the flow of group interaction. If distrust colors the environment, then distortions in communication will inevitably occur. Signs of distrust in the group are many:

- *General defensiveness:* Why should I reveal myself when nobody else here is doing it?

- *Attempts to change the attitudes and beliefs of others:* Everybody else here can be tolerant of abortion. Why can't you?

- *Persistent defense of one's public image:* I only seem distant here. I'm really quite different with my friends.

- *Attempts to make decisions for others:* Why don't you hug Marie and let bygones be bygones?

- *Avoidance of feeling:* It's what I think is important. My feelings come and go but my opinions are relatively stable.

- *Flattery:* You really know how to act in these groups. You do everything well. Have you been in one before?

- *Avoidance of conflict:* So John and I haven't worked things out. We can't work everything out here. Sometimes it's better to let sleeping dogs lie.

- *Advice giving:* You shouldn't let your mother dominate you like that.

- *Cynicism about what the group can accomplish:* We're in here only three hours a week and don't see one another outside. We can hardly expect miracles, so stop pushing.

- *Behavior that is based on strategy rather than spontaneity:* (The participant says to himself: If I don't join the others in confronting Peter, maybe he'll go easier on me.)

- *Impersonal talk:* People need people (instead of) I need you (or) Sit next to me.

An excellent sign of trust in the group is this: the participants say *inside* the group what they tend to say to wives, friends, husbands, and even fellow participants *outside* (where it is safer). If perceptions are shared and interactions worked out inside the group, then a climate of trust can be quickly established. Mistrust and fear make us ration out our communication in a most unproductive way.

It is true that misplaced trust can lead to deep experiences of shame. Still, a person can't wait to share himself until he is absolutely sure that he will be accepted by others. There is always the chance that one's gift of oneself will be spurned or go unnoticed, at least to a degree. However, the risk of laboratory training reflects the risk of life; too many of us fail to grow because we prefer a climate of absolute or excessive safety. The laboratory, because it is life in miniature (though it is also life under a magnifying glass), has supportive resources more readily at hand, for those who take risks and fail or are failed, than do real life situations.

Someone once called belief "prophetic of reality"; that is, if a person believes deeply enough in something, his faith will enable him to muster the forces needed to create that in which he believes. Trust, in the sense of entrusting oneself to others, can also be prophetic of reality: the person who dares to entrust himself to others goes far in creating a climate of trust in the group.

Consequent Support in Terms of Recognition. Studies have shown that a good deal of communication behavior is a function of the kinds of rewards a person receives. Rewards are reinforcing if they are related to my motives and needs. A communicator initiates communication when he expects some kind of reward on the basis of his past experience. Therefore, if the various forms of antecedent support outlined above are instrumental in getting group interaction started, it is consequent support in terms of encouragement and interaction reward that keeps it going and brings it to term.

Recognition and Appropriate Response as Encouragement. One meaning of trust is that: if I entrust myself to others (for example, by revealing myself to them), they will *respond* to me. Once a member of the group has participated in some form of interaction-enhancing behavior (expression of feeling, self-exploration), the other members should both (1) *recognize* (actively, behaviorally) the fact that he has acted responsibly and creatively, that he has done something good in the group, and (2) *respond* appropriately to his growthful behavior. Ideally, recognition and response merge into a single act. For instance, if a person engages in meaningful self-disclosure (story), it is not enough for the group simply to recognize verbally the fact that he has acted contractually—"You have engaged in story and this is a good thing in this group"—but they should react or respond to *what* he has said (if it is story, the assumption is that it is meaningful); that is, they should react personally to the disclosure. Appropriate response means response proportioned to the modality in which the other is speaking. For example, if A, perhaps only after screwing up his courage to take a responsible risk in the group, confronts B, then B's best response to A would be to examine himself on the issues suggested by A. Such a response would both recognize and reinforce A's behavior. Again, if A reveals himself significantly to the group, then B might respond by revealing himself along similar or relevant dimensions. B's act would indicate to A not only that he has listened carefully to him but also that he has felt a certain solidarity with him. Such an act would provide A with a good deal of reinforcement. On the other hand, if B were to reply to A's self-disclosure irresponsibly—for instance, by trying to upstage A with his own disclosure—then, obviously, his response would have the opposite effect.

The Necessity of Immediacy of Reinforcement. The experimental data on the timing of rewards suggest that rewards are most effective when they follow desired behavior almost immediately. Groups often fail in this regard. For instance, sometimes a participant will put himself on the line by disclosing something about himself that he thinks is important or takes the risk of trying to involve himself with someone he is afraid of, and, once he is finished, he is met with silence. I recall a group in which a young man risked himself, and, after he had finished, called for some feedback from the others. Luckily enough, this particular segment of the group experience was recorded on videotape. He received practically no immediate response from the others. In fact, when he called for response, most of the participants tried to "leave the scene" by bowing their heads or by looking off in a different direction. When I replayed the tape, I told them to watch what they did with their heads when they were asked to give some feedback. Then I asked the one who had put himself on the line how he felt at that moment. He said that he had felt alone, very much alone.

I would hypothesize that the more immediacy there is in reinforcement behavior in the encounter group, the greater the support experienced. This refers to temporal immediacy, certainly, for temporal immediacy of reinforcement means, at least, that the listener has picked up the cue "I am finished" and the cue "I would like some response." But it also refers to qualitative immediacy—that is, the degree to which the respondent really puts himself into the response. Indeed such qualitative immediacy seems more important than temporal immediacy.

   Cheap Empathy. Some sensitivity-training participants never miss an opportunity to give support, especially to the sufferer. They resemble professional wake-goers, their motto seeming to be: "I am always at your side (in time of disaster)." Such support is ritualistic, triggered by any sign of pain in the other, and seems to be directed toward fulfilling the needs of the one giving support rather than the one in need. Actually the one supported is seen as a stereotype—"one in pain," "one needing my support"—and to the degree that this is true, support is not authentic interaction with *this* person. On the other hand, oversupport might be a person's way of manifesting his own need for support or mothering. Support that is tender (not "mushy"), filled with feeling and emotion (though not sentimental), arising from the strength of the one giving support (rather than from his weakness) is most effective. The one who gives authentic support gives it in both adversity and joy, success and failure, because he does not see support as merely "propping the other up." Support means being "with" the other, especially at times of crisis. It takes both strength and skill to support the other because of the person he is and not because he is either victim or hero. Only those who are not afraid of universal contact with the other can provide a wide range of support.

   The St. Sebastian Syndrome. St. Sebastian was a Christian martyr who was killed, it is reported, by being shot full of arrows. This frequently happens in an analogous way in encounter groups. A person tells his story and, although it is vaguely sensed by others that support would be an appropriate response, no one knows how to go about the task of giving support. Being unskilled in the art of support, they tend to substitute a caricature. They begin to ask questions to show their "interest": "How do you feel? When did it happen? How are things now? How long has it been going on?"—and so on and on. This keeps the victim in the center of attention, of course, and does away with the need for real involvement or response. At first (at least this is my experience) the victim does not recognize the game; he thinks that the others are actually asking serious questions and he tries to answer them. Then he begins to feel that what is happening is either out of place, missing the point, or downright ludicrous, but, being polite, he still answers the questions for a while (though with less and less enthusiasm). His interlocutors keep pumping him with arrows (by this time even they are tiring of the game) until he and the interaction die. This caricature of support can also be called the "Is it bigger than a breadbox" syndrome both because of the Twenty Questions nature of the game and because such a question frequently seems as meaningful as those being asked. This does not mean that an occasional question cannot be both extremely insightful and deeply supportive, but it must be appropriate, nonritualistic, sincere, pithy, forceful, and a prelude to deeper involvement with the other. A question in this vein that is actually quite confronting can be more supportive than all the clichés and vapid questions put together. What about questions, then? In brief, if possible, skip the question and get to what's bothering you about the other.

Support versus "Red-Crossing." "Red-crossing" is a term that originated, I believe, at Synanon and means rushing to the aid of a group member like a Red Cross worker. Its connotations are obviously pejorative. Some people cannot stand seeing another in any kind of pain, physical or psychological, even in cases in which the pain is beneficial. For instance, if someone in the group is being confronted in a responsible way and therefore necessarily undergoing the pain associated with the process of confrontation, it is the red-crosser who comes to the aid of the confrontee in an effort to get him off the hook. He does so in a number of ways: he gives approval to the confrontee's behavior (whether the confrontee approves of *his own* behavior or not), he tries to rationalize away the other's guilt or responsibility, gives sp ˜hes the burden of which is that all of us are likewise sinne in general tries to show that the person being confronted is an innocent victim needlessly suffering.

> *John:* Henry, you run when anyone confronts you, no matter how much care he shows you.

> *Patricia:* Henry is a very sensitive person and we have to give him time to get used to confrontation. A lot of people here are afraid of being confronted. And we tend to pick on such little points.

The red-crosser is not at all like the person who intervenes when he feels that the confrontational process has become irresponsible, negative, or profitless in a particular case.

> *Tony:* I don't want to take Kathy off the hook, but we have spent a great deal of time with her. We are looking at her. Nobody is talking to anybody else. Sol and Tom have dropped out of the interaction entirely. How about letting Kathy assimilate some of this for awhile? How do you feel, Kathy?

This is not red-crossing but rather a simple recognition of what is happening in the group.

Silence as Failure in Support. "They also fail who only sit and wait"—to misuse a line from Milton. A group member once talked about the hurt she felt from the silent members. She did not feel they were hostile, but she found it difficult to engage in self-disclosure in front of people who willed to remain strangers. Even silence that is perceived as sympathetic is harmful if it is protracted.

A Final Note. No one can program the development of a climate of trust and support in an encounter group. The group, no matter how long it remains in existence, continually discovers new levels of trust and support. The initial reaction of many participants is often one of hopelessness: "I could never really entrust myself to you." However, as the group moves forward, even the most timid, encouraged by the risks the other members take, begin to move out into the group to establish and develop relationships with others.

While it is true that supporting another is a way of helping him, the purpose of support is not to make the other an "object of help." We are all subject to the human condition and as such need help and support from others. However, the support expected in the encounter group is the kind of support found in the interactions of people who are neither dependent nor isolationistic in their independence, but interdependent. Interdependent people help one another without making one another "objects of help."

Chapter 8

Confrontation

Confrontation is an important growth variable both in the laboratory and in life, but it is one that merits careful explanation for a number of reasons. First of all, confrontation in caricature is popularly taken as the symbol of laboratory training in general and sensitivity training in particular: "I don't have to attend a laboratory to tell people off and to give them their chance at me." It *is* true that some laboratory experiences are characterized by irresponsible confrontation, but this is certainly not generally true, nor is there any reason why it *has* to be the case in any given laboratory experience. Laboratories are designed to be growth experiences, not places where the psyche is laid open to possible destruction. And yet time and again people approach me asking me whether a particular individual should take part in a laboratory experience or whether a laboratory should be allowed to operate at all lest an individual or a group be exposed to psychic harm. Well-run laboratory experiences are no more dangerous than group therapy experiences, and I assume that the latter *are* run for the benefit of the participants. If confrontation is responsible—that is, proportioned to both the laboratory design and its population—then it will be a powerful force for growth.

Second, confrontation as a modality of mature human interaction has received practically no theoretical attention in the literature. Life itself without confrontation, however, is directionless, passive, and impotent. One of the reasons confrontation seems necessary for full human living is what might be called the bias nature of man: man, when unchallenged, tends to drift toward extremes; he becomes either too much "for" himself or too much "against" himself. Or—and perhaps this is the greatest tragedy—he merely drifts into the psychopathology of the average, which, given his potentialities, is also an extreme. The mature man is one who has learned to challenge himself and his own behavior; he is always looking for more productive ways to be and interact with others. But since he is really mature, when he fails to challenge himself he is grateful when his friends (and perhaps even his enemies) are concerned enough to confront him.

Confrontation has its place, first of all, in all mature human interaction. Because it is a modality of mature living, it also belongs both in interactions that attempt to explore human potentialities and deal with the psychopathology of the average (training laboratories) and in interactions designed to come to grips with more serious problems in living (psychotherapy). It is not strange to find confrontation in therapy and other interpersonal-growth experiences; rather, it is strange that normal men make such limited and ineffectual use of such a powerful growth variable.

Generally, confrontation takes place when one person (the confronter) either deliberately or inadvertently does something that causes or directs another person (the confrontee) to advert to, reflect upon, examine, question, or change some particular aspect of his behavior. In other words, some act on the part of the confronter—whether he is aware of it or not—acts as a stimulus to the confrontee, and this stimulus act has a specific effect on the confrontee: it challenges him, "pulls him up short," directs him to reflect upon or change some aspect of his behavior (behavior, that is, in the wide sense: overt acts, inaction, attitudes, moods). I believe that confrontation must be described or defined as generally as this if it is to include all behavior that is referred to in the literature as confrontational. Moreover, if it is defined this generally, it becomes quite easy to see that there are many different forms (both growthful and destructive) and many different degrees of confrontation. It is extremely important for participants to understand the nature of confrontation and to become acquainted with the different forms it can take, for it can be one of the most potent forms of interpersonal behavior, and its power should be respected.

At first glance, confrontation is a simple process, but it can become quite complex because of the variables involved: (1) the nature of the stimulus act, (2) the natural bias and motivation of the confronter, (3) the relationship between the confronter and the confrontee, (4) the manner in which the confrontation takes place, and (5) the effect the stimulus act has on the confrontee and the manner in which he responds to the confrontation. Each of these merits separate consideration.

The Forms of Confrontation

A wide variety of acts, both verbal and nonverbal, may have a confrontational effect on another. Group members, then, should become aware of the confrontational nature of what they do, for indeliberate (and thus uncontrolled) confrontation can be destructive. All the following acts can be confrontational: giving a person information he does not possess or possesses only inadequately, interpretation of another's behavior, directly challenging another's behavior, self-involvement of confronter with confrontee, group situational variables—for example, group exercises, being with strangers, or group self-criticism—withdrawal of reinforcement, and even the use of videotape. Each of these will be considered separately.

Confrontation through Information.   By the very fact that you stand outside of me, you have a view of me that I cannot possibly have. Therefore, everyone with whom I come in contact is, at least potentially, a valuable source of information about me. If confrontation were merely the transmission of correct and meaningful information by a concerned observer to a willing listener in order that the latter might engage in and grow through self-examination and subsequent behavioral change, then the whole process of confrontation would be simple indeed. In reality, however, such simplicity has to be learned; it is a goal rather than a starting point. The confronter might ask himself certain questions about the content of the confrontation. For instance:

Does the confrontee already possess the information? The confronter should weigh the consequences of telling someone what he already knows. For instance, if he says to another participant "You haven't said a word this evening, Frank," undoubtedly Frank already realizes this, and the confronter should do more than restate the obvious. For instance, he should explain how he feels about Frank's silence:

> Frank, I've found your silence up to now very annoying. I've just allowed myself to get annoyed without trying to find out what's happening to you. I should have gotten my feelings out much earlier.

This kind of statement is more honest, for it gets feelings out in the open.

Usually the confronter assumes that what he has to say is unknown to the other person in some sense. For instance, the confronter may think that the information, though known, bears repeating, that it should be brought into community, that repeating it here and now would have a specific effect, or that it needs to be dramatized. In this case, he assumes that the confrontee does not know the information well enough to act upon it; that is, the information is unknown in the sense that it either has made little impression on the confrontee or has not had any behavioral consequences. For instance:

> Mary, various people here have told you that you take practically no initiative to reach out and contact others. You know that well enough yourself. But I have no idea how you feel about your lack of initiative, whether you would like to be more active. I feel terribly frustrated.

Since the contract calls for initiative on the part of every member, the participants should challenge the nonparticipation of any given member.

How important is the information to the confrontee? "Your tie is crooked" is a relatively unimportant piece of information. "I notice that John never sits close to you" might be more important. "I find you rather cold and thoughtless" should be quite important. The objective importance of a piece of information will usually determine its impact on the group as a whole, including the confrontee, but the subjective meaning of any particular bit of information might differ greatly among confronter, confrontee, and the other members of the group. The confrontee might be quite sensitive to information that is relatively meaningless to the confronter and the group as a whole. For instance:

> You, being a rather large person, seem to have the joviality that I associate with larger people, but I wonder what your more serious side is like.

If the confrontee is very sensitive about his weight, this statement will have a different meaning for him. The intelligent and considerate confronter, then, is one who can judge not only the absolute importance of any piece of information but also its relative importance to the confrontee. Such a confronter might handle the confrontation in the example above a bit differently:

> *Syd:* Matt, you strike me almost as a stereotype of the heavy, jovial person. But I have a feeling that you might be sensitive about your weight.
>
> *Matt:* You're right. I am sensitive, but I knew it would be brought up eventually.
>
> *Syd:* I don't want to stereotype you. I've also felt that your joviality is a kind of mask or facade covering over some of the things that hurt inside. I mean the whole stereotype doesn't fit.
>
> *Matt:* You're right. Actually I'm a pretty frustrated, defensive person. But I can talk to you about it, Syd, because I feel you care about me, that you would never want to make fun of me.

Sensitive issues need not be avoided, but their sensitivity should be recognized.

Is the information fact or hypothesis? Facts deal with reality. Hypotheses are interpretations of reality. The following are confrontational statements that are factual:

> You don't look at me when you talk to me.

> You are too cynical for me. You are always saying what we can't get done instead of what we can do.

> We have so focused our attention on Trudy that everything else has stopped. I feel I can't speak unless I speak to Trudy. I feel everything is all clogged up.

When you do speak, you are very warm and understanding. I wish you would take more initiative here, because I think you have a great deal of humanity to offer.

Other confrontational statements can be highly interpretational:

You're afraid of me because you're afraid of your mother. You see a lot of your mother in me.

You have never successfully gone through the Oedipal stage. You haven't worked out the woman thing in your life at all.

You have been very nice to Sylvia in this group. It's been your way of getting my goat.

While interpretations need not be outlawed in the group, factual confrontations have greater validity. Interpretation too often degenerates into a game of logic or of "psychiatry," the purpose of the game being to avoid intimacy. Confrontation deals principally with behavior. The focus of interpretation is on the dynamics underlying behavior rather than on behavior itself. Dealing with the hypothetical sources of behavior might have some interest in itself but it seldom has any useful impact on behavior.

Confrontations that deal with the *why's* of a person's behavior are usually fruitless. Confrontations, even in the form of hypotheses, that deal with the *what's* and the *how's* of behavior are much more direct and concrete and tend to have a much greater impact on the one being confronted. "Your cynicism in the group is suffocating any kind of intimate interaction. I think John and Mary and I are scared to death of your wit." "I think you keep me at bay by always agreeing with me. Somehow your being too nice to me creates a distance between us." These confrontational statements, even though they are interpretational to a degree, deal with the *what's* and the *how's* of another's behavior. They are direct and concrete, designed to make the interaction more immediate.

The goodness of an interpretation is measured by its positive or growthful *impact* value: it must cause the confrontee to examine his concrete behavior, some aspect of his life style, in such a way that he is moved to modify this behavior in ways that improve his ability to involve himself with others. It is only the

empathic confronter who can come up with salient interpretations, for he alone is both outside and inside the confrontee
enough to put his finger on issues that are central to the confrontee's behavior and life style. Timing, too, is important: the good
confronter knows when the confrontee has opened, at least
enough to receive the full impact of an interpretation. Furthermore, confrontation must be related to the ongoing process of
the group and not just appear from nowhere; it should flow from
the group experience and be integrated into it. Finally, it should
be pithy, the starting point for interaction between the confronter
and the confrontee (and the rest of the group) and not a long-
winded statement with an air of finality. In general, long-winded
speeches in groups, no matter what their "pith and moment,"
tend to "lose the name of action."

The Direct Challenge. Although there is an element of challenge in any kind of deliberate confrontation, the explicit verbal
challenge can be a specific kind of stimulus act. In its simplest
form, the challenge is a suggestion, request, or demand that the
confrontee change his behavior in some way: "I don't deny that
it might be painful, but maybe it would be more growthful if you
were to try to involve yourself with us more," "Do you think
that you could honestly answer a few direct questions about our
relationship, yours and mine?," "Stop monopolizing the conversation." These are examples of bald challenges, whereas most challenges are situated in a much wider context of information and
explanation.

The direct challenge, as such, is neutral; that is, its growth
value for the confrontee (and for the group) depends on a number
of confronter, confrontee, and situation variables. For instance,
the confronter may act for a number of reasons: annoyance, concern for the confrontee, concern for the group or others influenced
by the behavior of the confrontee, or even a feeling that he should
say something. He may challenge the confrontee to do something
that is growthful or nongrowthful, relevant to the confrontee's
needs or irrelevant, possible or impossible; the confrontee may
or may not be prepared for the challenge: he may be hurt and
not ready to listen, he may be very anxious, there may or may
not be a climate of trust and support strong enough to sustain

the kind of challenge made. If the confronter fails to take these variables into consideration, he may end up blowing in the wind, much to his own frustration.

It is usually a mistake for the confronter to deal in demands rather than suggestions and requests when he challenges the behavior of others. Even if he does not intend to play God or omnipotent father, he may give the impression that he is doing so, and this tends to destroy any value that the confrontation might otherwise have. Since no one can predict with absolute certainty whether a change in another's behavior will benefit the other or not, there is always an element of hypothesis in every challenge. Therefore, as in confrontation through information, the degree of certitude associated with the hypothesis should dictate the manner in which the challenge is made. If the confrontee's behavior is obviously self-destructive or destructive of those around him, then the challenge to change can be put quite directly and forcefully. If, on the other hand, the confronter only suspects that a change in behavior will benefit the confrontee or others, the force or the demand element in the challenge must be proportioned to the certitude of the hypothesis underlying the challenge. For instance, whenever a confronter says to a confrontee: "You talk too much here," there are certain hypotheses and attitudes underlying such a challenge—for example, that the confrontee talks but says nothing, that he monopolizes the time of the group, that he is exhibitionistic, that he prevents other members who want to interact from interacting. Perhaps it would be more realistic to say that no matter how forcefully the confronter challenges the behavior of the confrontee, honesty demands that he deal openly with whatever is implicit in his challenge. If the confronter is willing to do this, he will soon learn whether he tends to challenge others because of his concern for them or because of his own needs.

Calling the Other's Game.   A special and quite effective form of direct challenge is "calling the other's game." According to Beier (1966), many people try to engage others in nongrowthful communication games. The game-player hides his real intentions in subtle cues rather than manifesting them in the overt message of the transaction. Most of us engage in these kinds of games at one time or another. We try to "engage" the person to whom we are speaking and still not reveal our own intentions. We get his attention and perhaps even his interest, but we do not tell him what we are looking for in the interaction. In therapy—and the same can be said of the encounter-group experience—Beier suggests that the therapist disengage himself from his client's game. The general instrument of disengagement is what he calls the "asocial response." Such a response is a refusal to give conventional replies to game-like interactions. It is a response that fails to reinforce the other's expectations. A few examples from Beier will illustrate what is meant:

Patient: I hate you.

Therapist: Go on [p. 51].

Patient: You are sure a quack. I don't think I should come to see you again.

Therapist: To seek help from an ignorant man like myself, this is crazy [p. 53].

A patient who claims she has been raped says: I like you very much, I even dream of you.

Therapist: You want me to lie with you on the couch? [p. 61].

Member 1: Nice weather today.

Member 2: Oh yes, I love a blue sky.

Therapist: And the clouds are just beautiful. And the wind is blowing so sweetly. And the sun is sparkling, bright and handsome.

Member 1: Let's get going [p. 149].

Patient: You only see me for the money.

Therapist: Why would anyone want to see a fellow like you for anything but money? [p. 60].

The asocial response disengages the game-player, thus placing him in a state of "beneficial uncertainty," a growthful kind of uncertainty because it takes place in an atmosphere of acceptance and support. Adler years ago suggested that one of the most effective therapeutic means is "spitting in the patient's soup." He can continue what he is doing but it no longer tastes so good.

What Beier proposes pertains, I suggest, not just to therapy but to encounter groups, other growth experiences, and to life itself. We would be better off if others were to respond to us asocially more often, catch us in our games, pull us up short. Beier in discussing family group therapy offers three guidelines for effective confrontation. He suggests that the therapist should intervene (1) when a member says or implies that someone else must change to solve his (the speaker's) problem ("If John would only stop drinking . . . "), (2) when any member makes a statement designed to maintain the past ("Well, this is the way my father was and his before him," "My wife wants me to be more aggressive, but I am what I am, and I can't help it"), and (3) when a member constricts communication by asking loaded questions or by giving connotations of which he may not be aware ("We always talk things over, don't we dear?" "Isn't that the way you've always treated me?"). These kinds of communications always crop up in encounter groups, and they should be challenged, but not just by the facilitator. The leadership of responsible confrontation should also be diffused among the members of the group. All response-restricting communications in the group should be challenged.

Involvement of Self as Confrontation. Not every kind of confrontation is explicit and direct. If a person tries to live up to the implicit contract of any encounter-group situation, he will take an active role in contacting others in various ways, especially through the modalities suggested here—self-disclosure, expression of feeling, and support. He will reach out and try to establish relationships of some closeness. But such contact is in itself confronting because intimacy is confronting. The person being contacted, the confrontee, feels himself being reached out to, he feels himself as the object of unconditional positive regard and empathy, he feels the impact of the other's congruence, he experiences the other as a locus of deep feelings and emotions. But in many ways he is probably not used to such behavior, especially such focused behavior, and it pulls him up short, it makes him take stock of himself, it makes him reach for responses that are not usually at hand. Intimacy, in our culture, is almost bound to have a confrontational effect, especially when it is set in sharp focus through the experiences of the laboratory. Since these indirect forms of confrontation can be as strong as or even stronger than more direct forms such as challenges of the other's behavior, the participants should be as aware as possible of the impact their behavior is having on the other. Some of the conflicts that arise in the laboratory stem from the fact that a participant's whole mode of behavior has been confrontational without his realizing it. For instance, the way a participant interacts with another may place the other under extreme emotional pressure. If this is the case, and it is not sensed by the confronter, then the confrontee or someone else should advert to what is happening: "John, you're making tremendous emotional demands of me right now, and I don't think I can reply to them at this time or under these conditions." Indirect forms of confrontation have the same effects, generally speaking, as more direct forms and should be pursued in the same way.

Group Situation Variables as Confrontational. Certain aspects of the laboratory experience itself are designed to have a confrontational effect on the participants. That a certain degree of intimacy is demanded in any encounter group is confronting enough, but the confrontational quality is compounded by the fact that the participant finds himself in a stranger group: he does not choose his bedfellows. As he looks around the group, it is as though each of the other participants were saying: "Here I am; you have to deal with me." It is not that any of the participants verbalizes or even conceptualizes such a challenge; rather, each *is* the challenge by his very being.

The verbal and nonverbal exercises used to stimulate communication in the group constitute another source of confrontation. In the "anonymous secret" exercise, the participants are asked to think of some secret that they would be unwilling to reveal to the group. Each person then in his own imagination pictures himself telling each of the members of the group. He tries to imagine whether he could tell it or not, how each member responds to him, how he feels about the response, and so on.

Then, without divulging the secret, people share with one another
their feelings. This exercise gets at the trust level of the group.
It is obviously confrontational because (1) it makes each participant think of some area of life he is unwilling to share and (2)
sometimes a participant has to say "I couldn't tell you because
I was afraid you'd ridicule me."

Processing   as   a   Group   Mode   of   Self-Confrontation.
Processing is a very useful form of group interaction. After
the participants have been interacting for a while, the interaction is stopped and the members process what has been taking
place—that is, they try to examine the nature of the interaction and answer such questions as "What have we been doing
here? How have we been interacting with one another? Who has
been interacting with whom? What have been our communication
blocks? Who has been active? Who has been silent? Which participants have assumed leadership roles? Who has been bored?" Processing gives the group an opportunity to be its own critic. The
members receive a kind of cultural permission to stand outside
themselves and act as critics of their own behavior. Remarks made
during this period are not taken as critical in a negative sense,
because responsible criticism is the very meaning of this processing interlude. Very often, because of the cultural permission, participants will find themselves saying things that they simply had
not been able to say during the group interaction itself: "We've
really been beating the air," "John, I think you wanted to get
us going, but you really monopolized the situation," "Bill, I'm
not so sure now that I was really honest with you," "I think we
showed how really afraid of self-disclosure we are," "I was just
too anxious to say anything." In the early stages of the group
life, processing is something that stands outside the regular
interaction of the group and thus gives the members an opportunity to realize that self-criticism can be quite constructive. Gradually, however, the members learn to incorporate processing into
the regular group interaction itself. Until such incorporation is
possible, processing periods provide excellent opportunities for
groups to confront themselves, to admit and frankly examine the
modes of flight that individuals and the group as a whole have
been using, and to do so without arousing intolerable anxiety.

Withdrawal of Reinforcement as Confrontational. Beier's asocial response is one that fails to reinforce certain behaviors that are seen as nongrowthful. Other such withdrawals of reinforcement in sensitivity groups can also serve as stimuli with confrontational effects. For instance, the person who uses humor to flee intimacy or to interrupt interactions that are uncomfortable for him is reinforced in such behavior when others laugh. Tension dissipates, and uncomfortable issues are sidetracked. A refusal to reinforce such behavior—that is, a refusal to laugh or to abandon an uncomfortable issue—is confrontational. Another example is the noninvolving monologue: if group participants tend to deliver monologues that prevent mutual interaction, then passive attentiveness will be an effective reinforcer. Confrontation in this case means interrupting and engaging the speaker. Part of the processing of group interaction should consist of pointing out the ways in which unproductive forms of behavior in the group are reinforced and in determining effective ways of withdrawing such reinforcement.

Videotape as a Vehicle of Confrontation. One of the primary advantages of feedback by videotape is its objectivity, its "cleanness." The confrontee is confronted by himself, and, since he cannot charge bias, rationalizations are relatively useless. Furthermore, the confrontee sees his behavior in context, he gets a view of the molar realities of interaction that ordinarily escape him. He can more readily see those aspects of his behavior that elicit "discrepant" feedback—that is, response to his behavior other than what he believed he would receive.

The Natural Bias and Motivation of the Confronter

A second factor to be noted in the confrontation process is the possible bias of the confronter. If the confronter, merely by the fact that he is not the other, by the fact that he stands outside the other, is in a metaphysical position to know the other in ways that are not available to the latter, it is also true that his separateness is a source of bias. The information that he feeds to the other is processed through the subjective filtering systems (the philosophical, value, and natural psychological systems) of the confronter and takes on the latter's color. Though this bias can be minimized (depending in large part on how close the confronter is to his own experiencing), it cannot be avoided altogether. Therefore, both confronter and confrontee should be aware of this phenomenon, for it is another source of possible error in the confrontational process. Even when the stimulus act does not consist of information, interpretations, or direct challenges on the part of the confronter but, rather, of the emotional impact he has on the confrontee, the confrontation may still be quite biased. The confronter's emotions are real, but they are not necessarily realistic. He may confront by expressing anger toward someone who really did nothing to provoke it (that is, he may be projecting the anger he feels toward himself). In such cases his emotions are biased (in some sense of the word), and thus the confrontee cannot be expected to respond as if he had actively provoked these emotions. In general, confronter bias should be controlled by group members other than the confrontee, for this latter should be as open as possible to constructive confrontation. But this very openness makes him less sensitive to sources of confrontational error. The others, therefore, are in a better position to pick up elements of bias and deal with them openly. This is one of the reasons why other members should "own" the interactions that take place between any two participants. Others can help surface both bias and defensiveness in confrontational dialogue.

An analysis of the motivation underlying confrontation has two distinct dimensions: (a) the purpose or function of an act of confrontation in itself and (b) the motivation of the confronter. Ideally, a high degree of correlation will exist between these two dimensions; that is, the confronter will choose to confront because of the growth functions he sees inherent in an act of responsible confrontation. These two dimensions, however, are separable, and the confronter can choose to confront out of motives that are less than ideal.

The Purpose of an Act of Confrontation. In general, confrontation is just one more modality of interpersonal contact and as such stems, ideally, from a desire on the part of the confronter to involve himself more deeply with the confrontee. Confrontation, then, is another kind of interaction that serves the general procedural goal of the group—to establish and develop relationships of some closeness.

The term "confrontation," when used in the context of international politics, connotes a standoff, a refusal to yield, an impasse, an irreducible separateness on the part of the nations involved. However, "confrontation" as used in the context of growth experiences has the opposite connotation: an act of constructive confrontation is an attempt on the part of one person to involve himself with another, a way of expressing his concern, a way of showing the confrontee that he is "for" him. Confrontation is a way of being with, rather than being against, the other, even though the confronter might disagree strongly with the confrontee. In training groups, then, the rule is simple: don't confront if you don't intend to involve yourself with the other.

> *Ted*: I don't like you, Steve.
>
> *Steve*: Why? What have I done?
>
> *Ted*: I don't know and I don't want to discuss it.

The person who confronts in such a way that he either does not want or fails to exhibit concern or a willingness to become involved with the confrontee is actually an intruder. In the example above, Ted is an intruder. Mere intrusive presence is inimical to interpersonal growth and is to be avoided in the laboratory group.

The direct purpose of an act of confrontation is not to change the behavior of the other but to create a situation in which it becomes possible for *him* to change his behavior. Confrontation is an invitation to the other to engage in self-examination. The confronter invites the confrontee into community to reflect on his behavior—whether the community be a dyad or a larger group. The purpose of confrontation is not to restrict the other but to free him. The confrontee is given an opportunity—in an atmosphere of trust and security—to step back from his behavior in order to see it in a different light and from a different viewpoint —that is, as it strikes others.

> *Sue*: Gene, I didn't like your behavior when we did the blind mill-around. You're too fresh.

Gene: That's your problem Sue. I enjoy being fresh.

Art: Tina, you try to dominate everyone here. You control everyone and don't let anyone get close to you.

Tina: I don't see you getting terribly involved. You haven't opened up any more than anyone else.

Neither Gene nor Tina accept the challenge to examine and possibly change their behavior. Gene wards off confrontation by defending himself. Tina wards it off by attacking her confronter. One purpose of confrontation is to bring the confrontee into more direct contact with his own behavior so that he himself might examine its value in the community of the group.

Another way of looking at confrontation is to see it as an instrument that might help the other reduce the amount of "cognitive dissonance" in his life. One way of reducing dissonance is to tailor reality to your own inner needs. For instance, there is a strong tendency to assume that the people one ranks highest on a scale of preference return the compliment, however unwarranted the assumption might be. Now, if I think that you have more interest in me than you really do, I may try to ignore whatever cues you emit that would indicate disinterest. There is an exercise that can be used in groups to get at this dissonance.

Exercise: Have subject "A" take a position of distance from "B" that he feels is appropriate ("A" might sit very close to "B"). Then have "B" do the same thing ("B" might move quite far from "A"). Then have the participants discuss what the exercise means for them.

In a group in which one of the goals is to try to establish and develop relationships of some closeness with others, the participants should let one another know where they stand with respect to one another. Concerned mutual confrontation, then, though it is sometimes painful, is a powerful tool for the reduction of dissonance. Responsible confrontation goes far in making reality the measure of one's thinking instead of the reverse.

The Motives of the Confronter. Ideally, the confronter engages in confrontation for the reasons listed above: he involves himself with the confrontee so that they might grow together. However, he may engage in confrontation for more idiosyncratic and less growthful reasons: to relieve his boredom, to ward off confrontation of himself ("the best defense is a good offense"), to punish the confrontee or the entire group, to take flight by engaging the group in game behavior, to relieve his own frustration and anxiety, to fulfill a need to dominate. Moreover, his motivation might be mixed; that is, he might confront for a variety of reasons, both good and bad.

> *Bob:* Tom, whenever you start talking I get turned off.
>
> *Jean:* I'm not turned off, Bob. And I don't think the other are.
>
> *Bob:* I guess I'm still hung up on your being an officer in the army, Tom. I hated the military. Now it's rubbing off on you.

Bob's motivation is obviously mixed. It would be futile for the confronter himself or the group as a whole to try to unravel the confronter's entire motivational skein, but it is useful for the participants to realize that confrontational behavior, like most human behavior, is multi-determined, multi-motivated. If, as far as the confronter is able to determine, his personal motives are in general agreement with the purpose of confrontation in the group, then he need not worry about ancillary motives. However, if he or the group suspects that ancillary motives are quite important or even predominant, then high visibility demands that he qualify his confrontation ("I may well be biased in my remarks because John and I have never gotten along") or, if he fails to do so, that his remarks be challenged by the other participants.

The Relationship Between the Confronter and the Confrontee

Both the long-range and the *ad hoc* aspects of this relationship affect the quality of the confrontational process. Confronter and confrontee might love, hate, or be indifferent or neutral toward each other, or the relationship might be a confused or muddled one, marked by an admixture of feelings, conscious and semiconscious, strong and weak, positive and negative. Moreover, such feelings may be mutual or one-sided. But even if the confronter loves the confrontee, at the moment of confrontation he might be acting from some lesser motive such as pique, jealousy, or momentary irritation. On the other hand, a confronter who generally dislikes another might rise above his feelings and engage in an act of concerned, growthful confrontation. In the encounter group, the participants are asked to be as aware as possible of their relationship to those they confront, including the quality of the relationship at the moment of confrontation. If the confronter has really done nothing to establish a relationship between himself and the confrontee, or if he has even rejected overtures of friendship on the part of the confrontee, this is obviously going to affect the dynamics of confrontation. Some research evidence corroborates what seems to be a common-sense observation: the behavior of liked persons is seen in a more favorable light than the behavior of disliked persons, even when the favorable perception is not accurate. The responsible participant will first of all be aware of the quality of his relationships and then try to rectify whatever bias might exist in his perceptions. An ideal atmosphere for confrontation is one in which the quality of the relationships is highly visible. This means a group culture in which members know where they stand in relation to one another because their relationships have been dealt with in community. It is evident, then, that confrontations that take place early in the group will often center around clarifying relationships ("We have really never spoken to each other, so I'm not sure where I stand"). The more real the relationships within the group are—that is, the more frank, open, and motivated by concern they are—the greater the chance of eliminating some of the natural-bias factors mentioned.

The Manner of Confrontation

Confrontation implies some kind of separateness, some kind of gap between confronter and confrontee that the former desires or feels impelled to close. His reasons for doing so, as we have seen, can either enhance or vitiate the act of confrontation. However, good motivation is not enough, for even the well-motivated confronter may confront in such a way as to thwart the desirable effects of confrontation.

Punishment as a Dimension of Confrontation.   This question is a thorny one, one that does not yield to ready answers. But it must be dealt with, because every act of confrontation, however responsible, seems to have a punitive dimension. Some people confront in order to punish and succeed in doing so, but, on the other hand, an act of confrontation not intended to be punitive may have quite punitive effects. This punitive tendency is a problem because research has not yet given us clear-cut answers about the growth value of punishment in human interactional situations. Research so far has shown that punishment, depending on the conditions under which it is administered, can have both positive and negative effects, but in practice punishment most often has negative effects, for few people know how to handle it well.

On the other hand, there is a certain amount of compelling evidence that even quite punitive confrontation can be helpful. In the encounter groups that take place at addict rehabilitation centers such as Daytop Village, there is a very great deal of extremely punitive confrontation. Yet it seems to be effective. It may be that addicts who go to these centers realize they are going to be put under extreme social pressure to change their life styles. They want to change, so they buy the whole package and in doing so "contract" to quite punitive confrontation. Furthermore, those who have witnessed these encounters see in this brutal honesty a kind of respect, a strong undercurrent of very real care. I am not suggesting that "brutality therapy" become a part of laboratories in interpersonal relationships. Rather I am trying to put my remarks about confrontation and the cautions I suggest in context. As a general rule, confrontation will be more effective if it is positive and constructive rather than negative, if the confronter confronts the strengths of the other and not just his weaknesses.

Proportion as a Key to Growthful Encounter.   The question here is: how can one strip confrontation of its tendencies to attack and destroy without stripping it of its impact? An attempt will be made here to formulate certain hypotheses about the manner in which confrontation must take place if it is to be a growthful rather than a neutral or even destructive experience. These remarks take on the nature of hypotheses because they are based on evidence that is principally observational and has not been confirmed by experimentation. The evidence so far suggests that the *proportional* nature of confrontation is most important; that is, confrontation, if it is to be effective, must be proportioned to a number of confronter, confrontee, and situational variables. In other words, the interaction of confrontation variables is as important as the variables themselves. It is assumed here that it is the socially intelligent person who can perceive the importance of these interactions and act upon them. In fact, this *is* one definition of social intelligence.

The strength of any confrontation arises principally from two variables: the sensitivity, or closeness to core, of the subject matter of the confrontation (for example, under ordinary circumstances the area of sex would be a more sensitive one, closer to the core of the confrontee, than, say, personal neatness) and the vehemence with which the confrontation is delivered (the Daytop Village approach would score very high on a scale of vehemence). These variables are additive, so that vehement confrontation in a highly sensitive area would represent the strongest kind of confrontation. If confrontation is to be responsible, its strength must be proportioned to a number of variables, among which are (1) the quality of the relationship between confronter and confrontee, (2) the current psychological state of the confrontee, (3) the possible disorganization the confrontee will undergo as a result of the confrontation, (4) the limits of the confrontee's capabilities, (5) certain group conditions, and (6) the contract, expressed or implied, that governs the group experience.

1. The quality of the relationship between confronter and confrontee. The confronter must ask himself: "What can the relationship between you and me bear?" To use metaphorical language, the more solid, the healthier, the more substantial the relationship is, the more powerful the confrontation may be (other conditions being equal).

Chuck: You are extremely low key, Al. When you speak, your voice is flat. You just don't seem to manifest any interest in anyone. I don't think you can complain about others' not responding to you until you give them something more to respond to.

Al: That's the picture I got of myself when I watched the videotape replay. You're right, Chuck. Hearing it from you somehow isn't so bad, because I know you're on my side.

If the confronter has tried to establish and develop a relationship of some closeness with the other, then it is more likely that strong confrontation will be effective, especially if the confrontee has reciprocated and taken some initiative to establish a relationship with the confronter. If the two have been dealing with one another in a context of mutual support, if each knows that the other is "for" him, then confrontation should be expected as part of their exchange.

2. *The current psychological state of the confrontee.* The strength of the confrontation must be proportioned to the current ability of the confrontee to support and act upon the confrontation. This means a number of things. If, for instance, the confrontee is already laboring under a good deal of anxiety, his immediate need might be for encouragement and support, and the confronter should ask himself whether it would be fruitful to confront him at this time (timing is always an important concern) or in this area of sensitivity or with this degree of vehemence. No participant is expected to be free of anxiety, but the level of anxiety should stimulate rather than paralyze. Confrontation should also be proportioned to the other's ability to change and, to some extent, his desire to change. One of the purposes of confrontation is to try to show the other that he does have the personal or community resources necessary to change an unproductive or destructive style of life. But if the confrontee is relatively eager to change, then he can ordinarily tolerate much stronger confrontation than a person who is not convinced that he can or should change. If the confronter is interested in the other, he will not mount a strong frontal attack when he knows that it has little or no chance of being effective.

> *Norma*: Louise, everyone's on your back right now. I'm getting the feeling that you want to say "To hell with all of you." I'm for letting this sink in a bit, but it is something I think you should deal with in this group. You help me deal with my garbage and I'll help you.

The confronter who, knowing that the confrontee has little or no interest in change, still hits him with an extremely strong confrontation is dealing in a form of shock therapy ("kick the TV and see what happens"). I have seen such tactics work, but they seem to be a last resort and effective only in the hands of an extremely socially intelligent person.

Some people, without being masochistic, are looking for strong medicine in terms of confrontation and are quite ready to respond favorably to it. I would hypothesize that this is especially true for many who voluntarily attend encounter groups, particularly if they have an accurate picture of what such a laboratory entails. Others, again without being masochistic, welcome even the punitive aspects of confrontation, for, to use an analogy from religious experience, they see in it an element of cleansing or expiation. On the other hand, it is possible that a person might seek confrontation precisely to prove to himself his own lack of self worth—"I know I'm a lousy person; tell me just how bad I am."

3. The risk of disorganization. Effective confrontation will usually induce some degree of disorganization in the confrontee, for it does touch his defense mechanisms. The responsible participant neither throws his defenses to the winds nor maintains them rigidly. The responsible confronter neither launches out to tear down the defense structure of the other nor hesitates to question defenses that keep the other from growing. The strength of the confrontation should be proportioned to the degree of disorganization foreseen and the ability of the confrontee, in the community of the group, to handle this disorganization. Confrontation means in some way challenging the unrealities of the world of the other.

Linda (weeping): It's awful being confronted with the fact that I'm dishonest in my dealings with others, that I do my best to manipulate and control people here.

Andy: I'm glad that you can admit it, even though it shakes you up. Now maybe you can do something about it, at least with some help from your friends.

Some have suggested that in general we tend to picture others as too delicate, too fragile, too incapable of handling strong confrontation and its disorganizing effects, and so we have been afraid of using confrontation. Or we fail to confront because *we* are afraid of the consequences for ourselves, because if we confront, we open ourselves to confrontation. Confrontation, then, is a two-edged sword—it cuts both ways.

4. *The limits of the confrontee's capabilities.* The encounter group should be diagnostic in the best sense of the word; that is, the participants should get a feel for one another's areas of potential and of deficit. While the participants should allow one another a wide latitude for growth, they should also come to a realistic understanding of one another's limits. Confrontation should be proportioned to these limits. Certainly the object of confrontation is to help the confrontee to move beyond his present limits, but even concerned confrontation cannot create potential where it does not exist. It may well be that the confrontee is not capable of the interpersonal styles that are preferred by the confronter, and it is a mistake to use confrontation in an attempt to get the confrontee to assume a style of relating that is not consonant with his capabilities. I have seen individuals and even groups spend unreasonable amounts of time trying to get the confrontee to assume an interpersonal style more similar to their own. This is a waste of time and also a way of avoiding one's own problems.

5. *Group conditions.* Confrontation must be proportioned to certain group variables; that is, the group must create a climate in which confrontation is viable. For instance, if the participants have achieved a certain degree of mutual trust and support, then relatively strong confrontation can be handled, even if the confronter himself fails to provide proportional support, for the other members will supply his deficiency. Confrontation is also much more possible in a group that has developed an open culture—that is, a group that has refused to tolerate the tacit understandings that tend to constrict or eliminate effective communication. In a group in which the participants show they are willing both to confront and to be confronted, confrontation soon becomes a constructive dimension of the group culture.

Adam: I've come to see that I am a rather stingy, tight-fisted person. I mean in the literal sense. I hold onto money even when I know I should be liberal with those I love. I wonder if that appears here in any way.

Jennie: That strikes some kind of chord. I see you a bit stingy with your affection, your care. It's as if you had only a little bit and are afraid of giving it away.

Adam: Yeah, that's the kind of thing I mean. Do any of the rest of you see me that way?

In a climate of trust, confrontation can be extremely beneficial, especially if the participants confront one another with respect to their unused potential.

6. *Contract, implicit or expressed.* It was suggested earlier that every group operates on some contract, whether expressed or implied. The trouble with implicit contracts in laboratory-learning situations is that they may or may not effectively provide for confrontation, and, even if they do, the conditions that regulate its use always remain vague. It seems only reasonable to assume that making confrontation one of the specific provisions of an overt group contract goes far to prepare the group for this variable. Through the contract, confrontation and other variables receive a high degree of legitimacy.

Ingrid: I don't want to hurt you, Norm.

Norm: You don't have to apologize for confronting me, Ingrid. That's one of the things we're here for. I take what you said in the caring spirit in which it was offered. I hope I don't have to apologize every time I confront you.

The hypothesis is that contractual legitimation of confrontation will increase its frequency, responsibility, acceptance, and effectiveness.

Growthful Response to Confrontation

Up to now we've been dealing with what the confronter should do in order to make confrontation a growthful interactional process. We now turn to the one who is confronted and what he should do in response to objectively growthful confrontation. Most of us naturally tend to *react* to confrontation, autonomically and emotionally, but few of us develop the art of *responding* to confrontation. The contract group is a laboratory in which this art is to be learned.

The most common reactions to confrontation are various forms of defensiveness ("My family has always been this way," "But I'm like this only in *this* group") and counterattack ("You haven't put out much in the group yourself"). Motivation, then, is crucial to the confrontational process. If the participant wants to change, he will want to be confronted and be willing to endure the unpleasant dimensions of confrontation. The ideal is that the confrontee enter actively into the confrontational process, that he become an agent in a dialogic process rather than just a patient suffering through something that is for his own good. Now we will try to clarify such confrontee agency.

Accepting the Invitation of Self-Examination. If growth-provoking confrontation is, in one sense, an invitation to self-examination as a prelude to possible behavioral change, then actual self-examination on the issues in question is the proper response to confrontation. The kind of self-exploration desired is one geared to the examination of behavior and focused on behavioral change.

> *Rich*: You can't let anyone suffer, Kyle. As soon as anyone gets some confrontation here, you're right there with a bandaid. It's unfair. You don't give anyone a chance to work things through.

> *Kyle*: I don't want to be seen like the big brother or the nurse. But what you say is true. I've done that to John, to Clara, and to you—at least. I feel so vulnerable myself. I guess I feel like I need a nurse at my side if I open myself up enough to be confronted.

Kyle does not defend himself nor does he attack Rich. Although it is painful, he explores his behavior. Too many people equate self-examination with a kind of amateur psychoanalytic self-exploration. The person who sees himself only through a maze (or haze) of psychodynamic interpretations can too easily become convinced that responsibility for what he does lies either outside himself or in hidden layers of the personality that are almost impossible to fathom.

Getting a Feel for How One Is Experienced by Others. The participant cannot respond to confrontation unless he is willing to venture outside himself. One of the most effective ways for him to learn is to drop his defensiveness, at least partially and temporarily, and try to understand the way in which he is experienced by the others in the group. Ideally, the confrontee—admittedly under somewhat adverse conditions—tries to become as accurately empathetic as he can during the confrontational encounter; that is, he tries to get inside the world of the confronter and actually feel how he is being experienced by the other. An exercise in confrontation that is useful at the end of a first meeting of a group is one called "first impressions." The participants merely exchange first impressions. As the group moves on, it is interesting to see how the opinions change.

Self-Confrontation. The contract calls for self-exploration in community as the response to confrontation, but such self-exploration may be either self-initiated or undertaken in response to a challenge by another. If the question is looked at a bit abstractly, a certain gradation appears: it is good for a participant to respond appropriately to unsolicited confrontation, but it may even be better if he actively seeks it—that is, if he manifests to the other members that he is open to meaningful hypotheses about himself and his behavior. It might even be best if the participant were to confront himself.

> *Ned*: I feel that I've been coming off rather punitively in our interactions here. Especially to you, Natalie. I don't like it. Help me deal with it.

Self-confrontation in the community of the group indicates a high degree of initiative and responsibility. The self-confronter realizes that he needs the group, both for support and for a corrective view of his behavior. There are, however, counterfeits of self-confrontation: some participants "get" themselves first so that the other members will not "get" them in areas more sensitive and more in need of attention. They want to avoid certain issues, so they take the initiative in less sensitive areas. Since many groups fall for this ploy, it would seem that a healthy confrontation culture would call for a combination of self-initiated and other-initiated confrontations.

An Openness to Temporary Disorganization.   The confrontee should be ready to accept (in some sense of the term) the disorganization induced by effective confrontation. Confrontation almost always demands some kind of reorganization of perceptions, attitudes, and feelings, and such reorganization is impossible without some kind of uprooting. However, if the participant is prepared for this disorganization, then his chances of responding, rather than merely reacting, to confrontation are greatly increased.

> *Rita*: I feel your concern and your warmth, Jude, and frankly they frighten me, they make me want to turn you off and run away, but that's precisely why I feel that I should take a good look at what's going on inside me, what's going on between you and me.

If the group culture itself and the participants individually provide support proportioned to the degree of disorganization, the possibility of successful confrontational interaction is further enhanced. Confrontation without support is disastrous; support without confrontation is anemic.

Confrontation and Conflict.   Even though the correct response to confrontation is self-examination in view of possible behavioral change, conflict and differences of opinion are not eliminated. Once the confrontee has assimilated the point of view of the confronter and has examined himself on the issues proposed, then together they might examine both areas of agreement and areas of conflict.

> *Todd*: I feel that you are afraid of me, Betty.
>
> *Betty*: I know I readily admit that I'm afraid of many men, but I don't feel I'm afraid of you.
>
> *Todd*: You do avoid me at times here.
>
> *Betty*: I know I avoid you. I'm afraid of men whom I see as strong. At least that's what I've thought all the way along. But in this group I've found out what kind of strong men I'm afraid of. I'm afraid of those who are strong in the sense that they are

warm, open, accepting, and willing to express this. You are strong in different ways. You can confront. You are very sure of yourself. But I don't feel your warmth that much. So I don't feel very afraid of you.

Conflict differs from defensiveness, attack, hostility, and aggression. It may be intense, but it is constructive rather than argumentative. In the example above, Todd probably learned something about himself and his "strength" that he had not thought of before. Conflict is a challenge, for those who participate in it, to accept one another's otherness. Research shows that conflict, if accepted and dealt with, can open up new possibilities for a group.

Some Suggested Rules for Confrontation

In summary, the following rules may help to make confrontation in the group a constructive process:

1. Confront in order to manifest your concern for the other.

2. Make confrontation a way of becoming involved with the other.

3. Before confronting, become aware of your bias either for or against the confrontee. Don't refrain from confrontation because you are for him or use confrontation as a means of punishment, revenge, or domination because you are against him. Tell him of your bias from the outset.

4. Before confronting the other, try to understand the relationship that exists between you and him, and try to proportion your confrontation to what the relationship will bear.

5. Before confronting, try to take into consideration the possible punitive side effects of your confrontation.

6. Try to be sure that the strength or vehemence of your confrontation and the areas of sensitivity you deal with are proportioned to the needs, sensitivities, and capabilities of the confrontee.

7. Confront behavior primarily; be slow to confront motivation.

8. Confront clearly: indicate what is fact, what is feeling, and what is hypothesis. Don't state interpretations as facts. Don't

engage in constant or long-winded interpretations of the behavior of others.

9. Remember that much of your behavior in the group, such as not talking to others, or expressing a particular emotion, can have confrontational effects.

10. Be willing to confront yourself honestly in the group.

No set of rules will provide assurance that confrontation will always be a growthful process in the sensitivity-training group. But groups can learn much from both the use and abuse of confrontation.

Chapter 9

Flight Behavior

It is not easy to engage in the kinds of interaction outlined in the previous chapters. Therefore, human nature being what it is, there is a natural tendency to find ways either to resist or to flee the work of the group. The message of this chapter is simple: Don't flee! Even though you engage in an encounter group because you want to, you will probably still resist the process to a certain degree, because it is anxiety-arousing and demanding. The group "threatens" you with both self-knowledge and intimacy, and your defenses rise to the challenge. Flight tendencies appear whenever the human organism is threatened by the often painful processes associated with personal and interpersonal growth.

This chapter is an attempt to "clip the wings" of the group member (or the group) in flight. It is a challenge to the participants to become aware of the principal kinds of flight behavior (many have been described in previous chapters) and to take a stance against flight by confronting themselves and one another when such behavior arises. Flight behavior in the group often mirrors the flight behavior in a participant's day-to-day life.

No attempt has been made to classify all possible flight modalities, for, given the ingeniousness of the human spirit in both devising modes of flight and disguising them, this would be an endless task. And since not only individuals but also entire groups may engage in flight, the following discussion is divided into two sections: the individual in flight and the group in flight. Even then there is some overlapping.

The Individual in Flight

The Ultimate Flight.   In the ultimate flight, the participant comes to the laboratory with no real intention of participating in it actively. Strangely enough, this happens more frequently than would be expected—perhaps especially in laboratory courses that offer academic credit. The person who comes unwillingly or as a spectator usually has a disastrous effect on the group. He is the cause of distrust, frustration, and anger. If the individual is coming unwillingly to the laboratory or if he is coming either out of mere curiosity or as a spectator, then he should not come at all. Passive, unwilling members corrode the group process.

Refusing Initiative.   Laboratories thrive on the initiative shown by the members. In the encounter group each member is to take the initiative to contact each other member in an attempt to establish and develop relationships with them. This refusal to take initiative is one of the principal ways in which individuals flee. They don't act in the group, they merely react. If there are several members like this, then the group stagnates. One of the most frequent failures to take initiative is the failure of the participants to "own" or get involved in the interactions of other group members. For instance, if Bill is talking to Jane, their conversation is not private. It belongs to the group, and if any group member has something to say either to Bill or to Jane or to both, he should speak up. This hardly means that everyone should constantly be barging into every conversation. But "barging in" is usually not the problem in groups. Most groups are too "polite." Everyone is allowed to "finish" his speech or dialogue before anyone else says anything. If the conversations between two or three people are so fragile that any interruption can destroy them, then perhaps the participants should work at making their interactions more robust. In good groups there is a great deal of spontaneous action. Sometimes there is so much interchange that the participants have to fight their way in.

Cynicism, Initial and Otherwise.   The participant who comes to the group experience with a closed mind and whose principal defense is a cynical attitude toward what is happening in the group will find it extremely difficult, if not impossible, to engage in the group interaction. Studies show that those who come to a laboratory ready to give themselves to the experience show positive results afterward, while those who expect to find the laboratory irrelevant find it irrelevant. Perhaps the most damaging failure of the cynic is that he refuses to initiate anything. To initiate is to show interest, and he cannot be caught showing interest in an enterprise that is in some way below him. He waits until someone else initiates something, and then he sits in judgment on what is happening.

> *Marlene*: Marge has had a pretty hard time of it this evening, Larry. You seem to be pretty unaffected.

> *Larry*: I can't get excited every time a female cries. It's OK if that's what she wants to do. It's not as if this were real life.

> *Bud*: You don't expect me to reform my whole life here, do you? After all, it's only a semester.

> *Sally*: Pat, you're pretty good at the kind of stuff demanded here. I'm not so good, but I guess I've got other talents. You win some, you lose some.

The usual posture of the cynic is one of silent judgment on the proceedings. He is all but invulnerable, for he can maintain the same posture when he becomes the object of confrontation. There are few defenses, if any, against ridicule, and the cynic, subtly more often than blatantly, ridicules or condemns what is happening. At times he may be swept along by the action of the group, and when the group situation calls for sincerity, he can even appear sincere. But he returns almost immediately to his former posture and hangs albatross-like around the neck of the group.

Silence. Some have tried to rationalize silence by claiming that there is no evidence that the silent person is not benefiting from the group experience; others say that the silent member is a point of rest in the group or that he is dynamic in the sense that the group must deal with him. There are a number of errors here. The silent person may well be *learning* something in the group, but he certainly is not *growing* interpersonally, for it is ludicrous to assume that interpersonal growth takes place without interpersonal exchange. Growth means, in part, that the person becomes an agent instead of a patient in his interpersonal contacts. Even if the silent member is considered dynamic in that he contributes to the dynamics of the group—his silence is a felt force, or he mobilizes the energies of the group, for they must deal with him—his silence is implicitly or explicitly manipulative and deleterious to interpersonal growth. The silent member frequently arouses the concern of the group ("Why is he silent?" "What is happening inside?"), creates feelings of guilt ("We have been neglecting him"), or provokes anger ("Why does he choose to remain an outsider?" "Why does he sit there in judgment of the rest of us?"). Thus, silence *does* manipulate, whether manipulation is the intention of the silent member or not.

A friend of mine told me that once during a forty-hour marathon, a girl sat against the wall for nearly fifteen hours. When the facilitator asked her what was troubling her, she said that she'd paid her money and was waiting for him to do something with her. But even in groups governed by a contract calling for initiative, some participants try to take refuge in silence.

Although quality of participation is, absolutely speaking, more important than quantity, there is a point at which lack of quantity is deleterious to the overall quality of an individual's participation. The marriage of quantity and quality of participation is something that must be learned in the actual give-and-take of group interaction.

The Interpretation-Insight Game.   Interpretation has already been discussed. Needless to say, the person who is constantly seeking psychodynamic interpretations of and insights into both his own behavior and the behavior of others is a person in flight.

> *Greg*: I see now! I fail to act because I still wonder what my parents will think. Even though my dad is dead.

The interpretation-insight game is a safe game, for it usually prevents the player from getting very close to any other member of the group.

Too many people try to solve their problems by insight alone instead of by the hard work that leads to behavioral change. Looking for new insights becomes a kind of way of life. Insight may, at least seemingly, reduce confusion in a person's life by wrapping up the conceptualizations he has about himself in a neater package, allowing him the illusory belief that he is "on top of his problems," for he can now explain his problems in rather sophisticated terms. But the person who hopes to be "saved" by insight is pursuing an illusion, hoping that the "ultimate insight" will be an answer for everything. Insight gives a person seeking help an out, a way to not deal with his behavior. Insight, in the sense in which it is under attack here, deals principally with cognitive systems and the relationships between cognitive systems. The problem is that cognitive systems and behavior are in separate compartments. The critics of insight might well paraphrase Kierkegaard's criticism of Hegel and say to the person struggling with his problems: "Your insights are beautiful castles; too bad you don't live in them."

However, there is one kind of insight that does make sense —the kind that results from *doing* something (rather than engaging in an exercise in logic).

Anne: This morning, Sally, when I told you that I thought that you were castrating Fred by treating him so harshly, my heart was going a mile a minute. Boy, did I learn how hard it is for me to challenge anyone! I had no idea of how much a peace-at-any-price kind of person I am.

This kind of insight is not a game but a starting point for behavioral change. If a person throws himself into the give-and-take of the interaction, he learns new things about himself and has other things confirmed. Fear of the emotion that arises from meeting others directly moves us to try to substitute intellectual games. They are poor substitutes.

Humor. Humor is a two-edged sword. For instance, it can be used to ease the punitive side effects of confrontation and thus facilitate interaction, but it can also be used as a weapon to destroy attempts at deepening the level of the interaction or broaching sensitive topics. When humor is used to dissipate tension, it does just that, but it does so without getting at the issues underlying the tension. Whenever either an individual or a group adopts humor as a consistent component of interactional style (for example, when the group spends five or ten minutes in banter at the beginning of a session, or when a member becomes humorous whenever a particularly sensitive issue is brought up), such behavior should be challenged. This is not to deny the value of humor in individuals and in groups, however. The humorless individual and the group that cannot laugh at its own incongruities are both dull.

The Questioner.   Something has already been said of questions in the section called the "St. Sebastian Syndrome." It is amazing how many participants handle emotional interactions by means of questions. For instance, John reveals himself at a fairly deep level and expresses a good deal of emotion in doing so. He ends by saying:

> So when anyone here shows me too much affection, I get a knot
> in my stomach, I perspire, and I want to jump out of my skin.

The others, unable to handle either John's self-revelation or his emotion, begin to ask him questions instead of supporting him or engaging in counter self-sharing.

> *Adele*: Have you always been this way, John?
>
> *Peter*: Do you think that has anything to do with the way your parents treated you?
>
> *Michele*: Is it the same whether a male or a female shows this affection to you?

Or the most awful question of them all:

> *Eve*: How do you feel right now?

Questions can be effective if they get at the guts of the interaction and also reveal something about the questioner in the process (for example, that he is concerned or anxious). But on the negative side, questions can be an attempt to intellectualize emotions expressed in the group or they can even be an expression of hostility, a way of picking at the other. Some people ask questions because they think that they should say *something* or interact in some way, and questions constitute the safest way of doing so. "Why" questions especially can lead the group on a wild-goose chase, for they precipitate interpretation-insight games. "What" and "how" questions, on the other hand can be quite effective. "What are you trying to do to me?" and "How do I annoy you?" are quite direct questions that add to the interaction instead of sidetracking it. One rule will help avoid flight-by-questions: before asking a question, see whether it can be turned into a statement. Instead of asking "How do you feel?" (a tremendously overworked question in encounter groups), express your own sentiments: "You seem very depressed right now and I feel a bit guilty because I think I'm one of the causes of your depression."

Rationalizations. Rationalization plays a part in almost every defense mechanism. In sensitivity groups, an almost infinite variety of rationalizations are available to the participant who finds that he is not giving himself to the group experience. As the proverb goes, he who cannot dance says that the yard is stony. It is possible to mention here only a few of the more commonly used rationalizations. One is the distant-fields syndrome: the participant says to himself, if not to others, that he would be doing well if he were only in that *other* group. It is the peculiar combination of personalities in *this* group that prevents him from getting on with the work. Once he has convinced himself that the obstacle is his *environment* (including the facilitator) rather than himself, he proportions his participation to his discovery. Sometimes it takes the form: "Even I cannot cope with this group."

Others project their own inadequacies on the laboratory experience itself: "This is a contrived situation, quite unreal; it does not facilitate real interpersonal contact." The fact is that no one claimed that the group situation was not contrived; the laboratory is both real and unreal, but the person who becomes preoccupied with its unrealities is in flight. It is too easy to blame these unrealities for one's own failures, just as it is easy, for the person who is afraid of failure or who feels he has failed, to become cynical about the powers of the group.

A rather silent member of a group once suggested an interesting rationalization for his nonparticipation: he said that he was participating *outside* the group but found it impossible to participate inside the group. Inside the group he was *learning*; outside the group he was *doing*. Other statements that summarize rationalization processes are: "No one else is doing anything," "I really don't know what is holding me back," "I really don't know what we are supposed to be doing here," "I'll let you know whatever you want; just ask me,"—and the list could go on and on.

Boredom. "Endured" boredom is a favorite form of flight for some. For instance, one of the participants witnesses an interaction that has little meaning, at least for him, but at that time he chooses to keep his disaffection to himself. Then, after perhaps an hour of enduring the interaction, the bored person, in a burst of "honesty," tells the group how bored he has been. At this point, what may appear to be honesty is, in reality, an act of censure and punishment and at the same time a confession of noninvolvement. If a participant has allowed himself to be bored for any length of time, he should not punish but apologize, for he has violated his obligation to take initiative and "own" the interactions taking place in the group. The participant who does nothing about his boredom deserves it.

Dealing in Generalities. One of the most common failings of participants is a generality approach to the interaction. When speaking, they use "you" and "one" and "people" instead of "I," they state and restate general principles without applying them to themselves or to others, and they address themselves to the group at large instead of contacting individual participants. Being specific *does* make a difference in the sense of immediacy in the group. There is a world of difference between the statement "You get scared to bring up what is really bothering you" and "I am really afraid to talk about the things that bother me here." It does make a difference when participants talk about specific incidents and people inside the group instead of speaking generally about the group culture. It does make a difference when a participant directs a high percentage of his remarks to individuals in the group instead of speaking generally to the whole group. When a participant addresses the whole group through the immediacy of his contact with another individual, the entire interaction becomes less abstractive and more engaging. In experimenting with groups, I have arbitrarily banned the use of "people" and of "you" and "one," refused to let others make general statements, and demanded that the participants address themselves to particular individuals instead of the entire group. The resulting culture, while somewhat artificial, at least initially, dramatizes the lack of immediacy that preceded it. Rid the group of the generalities that plague it and you rid it of one of the most common sources of boredom.

Low Tolerance for Conflict and Emotion. Inevitably there are some participants who have a low tolerance for conflict or strong emotion. When conflict and emotion run high, these participants react in one of two ways (or in both ways at different times): they either withdraw from the interaction or they try to stop what is taking place. In conflict or confrontational situations, they may become mediators, saviors, or "red-crossers." They defend the confrontee, chide the confronter, and in general pour oil on the waters of conflict. Nor is it just negative emotion that they find intolerable. Often strong positive emotions are just as threatening, and these, too, must be tempered by humor, changing the subject, and other ploys. This does not mean that no one should ever intervene in conflict situations; some conflict situations are unprofitable and forms of flight in themselves, and they deserve to be challenged. It does mean, however, that one should try to be aware of, and perhaps declare, his motivation when he does intervene in conflictual or other emotional situations ("intervene" is used here as distinct from "participate"). The participant who can declare his discomfort during even responsible conflict without trying to sabotage the interaction has taken the first step toward handling his low tolerance for emotion.

No Space to Move. One of the functions of growth experiences is to reveal to the participant the possibilities for growthful change. He can resist this revelation in a variety of ways. He may deny that any change, or at least the changes proposed, would be beneficial or even possible. That is, he may present himself as having no room in which to move. He sees himself as hemmed in by his heredity, environment, and history. Too often his response to confrontation is: "I can't do anything about it; that's the way I am." The problem here is not that the participant refuses to move in the directions suggested by others but that he refuses even to entertain the possibility of moving. He might refuse to do so in the name of personal freedom and integrity or from a sense of being the victim of forces outside his control, but responsible confrontation is not an attack upon a person's integrity nor a call to do the impossible. The person comes to the group not to be remade but to entertain possibilities for change. If he resists this goal, he is resisting the very *raison d' être* of the group. If a participant finds that his interactional style is heavily loaded with statements of defense or attack, he might well entertain the hypothesis that he is unwilling to examine even the possibility of change.

Control.   Some members flee deeper involvement with others by controlling what takes place in the group. They attempt this control in a number of ways: by focusing the attention of the group on themselves continually, by being overtalkative, by cynicism, by hostility, by silence. A devastating form of control is "pairing." Member A moves into a coalition with member B and perhaps member C and through this coalition they determine what takes place (or especially what is *not* to take place) in the group.

> *Rudy*: I think we're staying at safe distances from one another. Let's stop and process how "safe" we are playing it.
>
> *Rhoda*: Rudy, you're always worried that nothing is happening. Why don't you cool it and let what happens happen?
>
> *Nate*: We have to be natural here, Rudy. You want to go too far too fast.
>
> *Ellen*: I love to watch the dynamics of groups.

Rhoda, Nate, and Ellen team up and decide what is *not* going to take place in the group.

The Group in Flight

Obviously, some of the resistances and flight modalities outlined above could apply not only to individuals but to the group as a whole. This is especially true if the group establishes a culture in which a particular flight modality—for example, intellectualized interpretation—is tolerated or even encouraged. Certain other behaviors, however, indicate the resistance or the flight of the group as such, though here again, any particular behavior might characterize an individual rather than the group. Some common group-flight behaviors are outlined below.

Analysis of Past Interactions. Detailed analysis of past interactions in the group is both a caricature of processing and a seemingly popular form of flight. For instance, the group, in an afternoon meeting, discusses in detail the dynamics of the morning meeting. This discussion takes place under the guise of "working things out" and "getting things straight," but the hidden purpose is frequently to gain a respite from intimate interaction. The group analyzes not to put things in perspective but to put the interaction at a distance. Sometimes these historical analyses take place during the same meeting; that is, the participants interact for a while, and then its interchange degenerates into a "This happened and then that happened" or a "No, B got angry and *then* A got angry" situation in which the here-and-now is abandoned for a safer there-and-then.

> *Dominic*: This all started when Clem confronted Liz about her nonparticipation.
>
> *Liz*: That was only part of it. I really started when you said that you were bored.
>
> *Clem*: Then I suggested that the whole thing was boring because of disinterest and that is when I mentioned you, Liz.

This kind of thing could go on forever. An ounce of interaction is followed by a pound of analysis. In worthwhile processing, the participants become critics of the group culture and of their own interactions, with a view to improving both. Historical analyses, on the other hand, usually become ends in themselves; the group treads water, as it were, in order to avoid involvement. It is interesting to see what happens after a group concludes an historical analysis of its interactions; frequently an uncomfortable silence ensues, which is some indication of the irrelevance of the preceding analysis. One good sign that a group has degenerated into a useless kind of historical analysis is the increasing use of the past tense by the participants.

Irrelevant Serious Conversations.   Sometimes groups flee the work at hand by engaging in worthwhile discussions that really have no relation to the goals of the group. I once sat in on an encounter-group session and listened for a while to a rather high-level discussion of some of the most important social issues of the day. Had a total stranger come into the group without knowing its general purpose, I am sure he would have thought it was a social-action group that was taking itself seriously. Such conversation would have been most appropriate at another time in another place, but in the training group, it was merely an expression of discomfort and anxiety and a flight from the real purpose of the group. Usually, such serious discussions take place early in the life of the group, for most people assume that when they are talking intellectually about serious issues, they are in close contact with their fellow discussants.

Some participants are willing to talk about serious problems that are affecting their lives outside the group while remaining unwilling to make closer contact with their fellow participants. College students, for instance, are often quite willing to talk about their problematic relationships with their parents. Admittedly, this is a serious concern, but usually it is not a here-and-now concern for *this* group of people. Such parent-problem talk inevitably involves long descriptions of back-home incidents that cut down on group interaction. In some cases, the problems brought up can be made relevant to the group. In most cases, it leads to monologue and the deadening of group interaction.

Irrelevant serious conversations are usually difficult to challenge, for it seems ignoble to interrupt something that is so worthwhile in itself. I tend to interrupt such conversations, however, and have the group process the discussion itself. Soon the participants realize that they are avoiding the real issues before the group.

Turn-Taking and Dependence on Exercises. Some of the exercises commonly introduced into human-relations laboratories involve "turn-taking" or "going around." For instance, early in the life of the group, each member might be asked to give his first impressions of every other member. Turn-taking provides a structure that both forces each member to participate and provides a certain amount of institutionalized safety. As the group moves forward, however, there should be less and less reliance on such artificial methods of stimulating meaningful interaction and of getting everyone to interact. If the group goes dead, or if any particular participant is failing to interact, these issues should be dealt with directly without resorting to artificial communication stimulators. Sometimes a group has to live with its own inertness in order to discover that it has within itself the resources to deal with its own problems. Exercises can be used effectively to introduce different kinds of experiences into the group, to stimulate interaction, and to break through communication blocks; but if they are used excessively, they rob the group of its initiative. If the facilitator feels that he must make frequent use of exercises in order to keep the interaction going, his feeling is probably more a sign of his own flight from anxiety than of his interest in group communication. As the participants grow, they will resist artificiality and attempts to manipulate them into doing what they know they should do themselves.

Dealing with One. A common occurrence in groups is the phenomenon of one person's becoming the focal point of the group interaction for a relatively extended period of time. In its extreme form, this means that eventually each member has his turn as the focal point of the group (and this results in a kind of amplified version of "going around"). There is no particular reason that the group may not give extended attention to the problems of an individual member, but, in practice, the way in which such attention is given frequently involves elements of flight. Once a single individual becomes the focal point of the interaction, usually a number of members fall silent, becoming interested listeners because they have nothing to say to the participant in question, perhaps because they do not have the same problems. These silent members are in flight. No structure that significantly reduces the involvement of a number of members (or serves as an excuse for their noninvolvement) should be tolerated; if dealing with one individual has this effect, then it is not being handled properly.

It is possible, however, to deal with one member in a way that does not cut down on the involvement of the other participants. First of all, no participant should be allowed to excuse himself from the interaction simply because another participant is the focal point of the discussion. Second, dealing with one participant does not exclude interactions between other participants. Too often it is assumed that if the group is dealing with one, the only permissible interactions are those between the focal participant and someone else. The situation thus produced is artificial and stultifying. Third, the focal participant, himself realizing that there is a tendency on the part of some to withdraw from the interaction in a "dealing with one" situation, should take the initiative in bringing silent members back into the discussion. The problem is that most focal participants become somewhat passive and merely allow themselves to be dealt with. Another danger inherent in a "dealing with one" situation is the tendency to focus on there-and-then incidents and concerns rather than the here and now. This tendency serves the cause of alienation. The responsibility for maintaining a here-and-now culture in a "dealing with one" situation falls, to a great degree, on the shoulders of the focal participant. He cannot escape talking about there-and-then concerns, but it is his responsibility to make these concerns relevant to the here and now. For instance, it may be that the there-and-then limits his ability to involve himself with the other participants—that is, his past history affects his here-and-now participation and involvement; or, on the other hand, it may be that he has found it both easy and rewarding to involve himself with group members, a fact that stands in stark contrast to his there-and-then situations.

Bryan: My dad has a Ph. D. in physics. My sister graduated with top honors. They treated me like I'm not too bright.

Val: I don't see how that affects you here and now.

Bryan: It sure does, Val. I keep wondering if I'm living up to your expectations and I generally feel I'm failing.

Carol: You're not failing as far as I'm concerned, Bryan. But Val, I have somewhat of this feeling toward you. I wonder, not if I'm failing the group, but if I'm living up to your standards.

Bryan makes his there-and-then concerns relevant to the here-and-now. Carol goes one step further. She does not feel that she has to "finish" with Bryan. She feels free to deal with Val since it is in the spirit of what is happening in the group.

Sometimes when a person other than the focal participant brings up concerns of his own that are similar or relevant to those of the focal participant, he will be silenced and chided for not giving the other enough time. This defeats the very purpose of the group, which is not to deal with one another serially but rather to engage in mutual self-sharing. Giving the focal participant enough time ("I don't think we're finished with Ted") frequently means keeping him artificially in the limelight, making him a case, constituting him as a problem, which the group must in some way solve. Those who mother the focal participant, protecting his time and interests, should examine their motivation for doing so.

Personally, I prefer group cultures that do not make a habit of dealing with a single individual in an extended way. Even the most dramatic aspects of the lives of the participants can be dealt with in the give and take of a culture that calls for the more or less continued participation of all members. If a member reveals a dramatic dimension of his life, he needs support rather than time. To assume that this support can take place only by his becoming the central actor (or patient) on the stage for an extended period of time is unwarranted. One of the most effective forms of support is mutual self-sharing, for it is a way of saying "Thank you for trusting me, I trust you, too" or "Thank you for sharing yourself with me, it gives me the courage to share myself with you." It is also unwarranted to assume that the group can work completely through any participant's self-revelation at any one time. Such grandiosity covers a multitude of uncertainties and anxieties in groups. The group should feel free to return again and again to anything that has been said in the group and to look at new facets of self-disclosure in the light of subsequent group interaction.

Tacit Decisions.   Unchallenged modes of behavior tend to pass into the group culture as laws, which, when once made, are very difficult to change or abrogate. Tacit understandings can affect almost any aspect of group life: the content of discussions ("We don't talk about sex here"), procedure ("When one person is talking, no one should interrupt him and he should be given all the time he wants"), depth of interaction ("We really don't want to wash our dirty linen here" or "There are some things that we should just keep to ourselves"), rules ("Coming late or absenting oneself from this group is not an offense"), style ("Humor is allowed almost anytime during interaction"), goals ("Our purpose is to decrease the discomfort that we feel in being together with one another" or "Cooperation with one another is not a group value in this situation"). Since flight by tacit decision can take place very early in the life of the group, it is up to the facilitator to explain both the process involved and the consequences of such understandings and to challenge them before they gum up the proverbial works.

Ritual Behavior.   The encounter group can too easily take on a ritualistic atmosphere, devoid of eruptions of any kind, in which sameness soothes. Although the participants become quite comfortable in such a ritual, there is still the illusion that something serious and worthwhile is really taking place. In a ritualistic group culture, not only do the same issues come up over and over again, but they are handled in the same way. For instance, X's silence and general lack of involvement are dealt with from time to time, but little is done between ritualistic confrontations to bring X into the interaction. It is almost as though the participants were to say to themselves from time to time: "Since nothing in particular is taking place right now, we might as well make a group assault on X again." One way to dramatize the ritualization of the group is to replay a videotape of, let us say, session three, together with a tape of session eight. If the same issues are being dealt with in the same way—that is, if it would be impossible for a stranger to determine which was the earlier and which the later session—then it may well be that the group has ritualized itself as a form of resistance or flight. A ritualized group culture is a sterile thing, of which boredom and resistance to attending (reluctance to come, coming late, actual absences) are the inevitable signs.

Lowest-Common-Denominatorism. When even one person in a group displays indifference toward the goals of the group, the efficacy of the group is lowered. Often the encounter group moves along only as rapidly as the slowest member. The problem of the lagging, delinquent, or deviant member is one that arises naturally in groups. An effective contract can help control deviancy by eliminating the unmotivated (especially if the contract is freely chosen and not just imposed) and by eliminating the kinds of vagueness and ambiguity in group process that often engender the indifference or apathy of the deviant member. However, there is no ultimate way of ensuring the interest and cooperation of all members. The problem of lack of motivation is one of the most difficult to handle in all kinds of growth experiences. Perhaps the facilitator could discuss the possibility of the deviant member in the group before the group gets going. If the possibility is discussed, the delinquent member may have a less retarding effect on the group (that is, if some member does prove to be a deviant). Whether such a member should be expelled from the group or not is a moot question. He certainly should not be allowed to absorb the energies of the group.

> *Donna*: Reggie, when are you going to open up here?
>
> *Stan*: Donna, a number of people have shown a great deal of care toward Reggie, both in terms of warmth and in terms of caring confrontation. He has not really responded very much at all. I would rather use our energy differently. Those who really want to get involved are getting involved. My feeling about Reggie is: if he wants to join what is going on, he is welcome at any time. But I don't think we should mother him any more.

Reggie has to provide his own motivation. If the Reggies of the world drop out, that is their choice. The natural tendency of groups that lose a member who really does not want to be there in the first place is to spend a good deal of time dealing with their own feelings of guilt and loss. I'm not sure that this is time well spent.

What about members who absent themselves from the group, are chronic latecomers, or leave early? I think that deviant behavior should be dealt with in the contract that governs the interaction of the group. At the minimum the contract should call for the confrontation of the delinquent.

> *Bess*: I'm sorry I'm late. It looks bad for me, especially since I missed last week.
>
> *Edith*: Bess, we agreed to confront those who deviate from the rules all of us saw as helpful for running this group. I need something more than an "I'm sorry." It looks like this group is not very high on your list of priorities. I feel a bit hurt right now.
>
> *Bess*: The heart of it seems to be that I'm not the "committed" type. Not here, not anywhere. It's not that I've singled you out.
>
> *Vince:* That's one of the most honest things you've said here, Bess. I feel I'd like to see you experiment with being committed to this group and deal with the feelings that well up in you. And deal with your behavior.

The deviant member should not be allowed to destroy or seriously impair the work of the group. The common good comes first.

Flight versus Maintaining Adequate Defenses.   Laboratory-learning experiences, if carried out responsibly, give the participants a relatively safe opportunity to lower some of their defenses in the name of growth. No growth experience, however, should demand that the participant divest himself of his defenses entirely. But maintaining adequate defenses (even in the process of lowering them) and resistance to flight are two different processes. The more common danger is that the participant will not drop his defenses enough to allow the experience to have its impact on him. The person with crumbling defenses either refuses to participate in such experiences because he senses the danger or reveals his tenuous defense system behaviorally early in the life of the group. Selection procedures should screen out those with crumbling defenses so that the group culture can bring to bear fairly strong pressures on those with more than adequate defenses.

A Concluding Note. Don't flee! Instead turn back to the beginning of Chapter 3, to the rating sheet that begins that chapter. Instead of fleeing, do your best in the group to increase the quantity and the quality of the kinds of interactions and the ways of interacting outlined in that scale. Watch the other members of the group. Some will be quite good at some kinds of interaction (such as self-disclosure). Without being slavish, try in some way to imitate the good things you see in the group. Don't be afraid to use your own talents so that others can learn from you, too.

Remember, also, to keep what you are trying to do in the group in perspective. Life is certainly more than the group—it is poetry, art, music, religion, work, problem-solving, politics, involvement in the social order, vacations, fun, romance, learning, books, friends, pain, career, success, failure, the rest of the world, dying. What you learn in the group should enable you to involve yourself in the rest of life a bit more creatively.

Encounter Group Checklist

Use the following checklist to assess your participation in the group and the quality of the group interaction. This checklist will probably be more helpful after a few meetings when you get an experiential feeling for the contract.

1. *Tone.* What is the tone of the group (spontaneous, dead, cautious)?
2. *Commitment to goals.* Answer the following questions in view of the general procedural goal: that each member is to attempt to establish and develop a relationship of some intimacy with each of the other members.
   a. Are members working at establishing relationships with one another?
   b. Have a number of significant relationships emerged?
   c. Are relationships becoming deeper or remaining superficial?
   d. Is there consensus that the group is moving forward?
3. *Initiative*
   a. Do members actively reach out and contact one another or do they have to be pushed into it?
   b. Is there some risk-taking behavior in the group?
4. *A climate of immediacy*
   a. Do members deal with the here-and-now rather than the there-and-then?
   b. Are there a large number of one-to-one conversations as opposed to speeches to the group?

c. Is the content of interactions concrete and specific rather than general and abstract?

d. Do members use "I" when they *mean* I instead of substitutes (one, you)?

e. Do members avoid speaking for the group, using the pronoun "we"?

5. *Cooperation*

   a. *Cooperation.* Is there a climate of cooperation rather than one of antagonism, passivity or competition?

   b. *Polarizations.* Are there polarizations in the group that affect the quality of interactions (for example, leader versus members, active members versus passive members)?

   c. *Owning interactions.* Do members tend to "own" the dyadic interactions that take place in the group? When two members are having difficulty talking to each other, do other group members help them? Do those having difficulty seek the help of others?

   d. *Check it out.* Do the members, when they confront one another, check out their feelings and evaluations with other members?

   e. *Hostility.* Is there any degree of hostility in the group? Does the group or do individual members wallow in it or do they seek to resolve it? Is there covert hostility? If so, what is done to bring it out into the open?

6. *The principal modes of interaction*

   a. *Self-disclosure*

      (1) Was it appropriate, that is, geared to establishing here-and-now relationships of some intimacy?

      (2) Was then-and-there disclosure related to the here-and-now—that is, specifically to this group or one's relationship to this group?

      (3) Was it related to the process of encountering (establishing relationships) rather than counseling (dealing with there-and-then problems)?

      (4) Was it meaningful disclosure or superficial?

   b. *Expression of feeling*

      (1) Did people deal with feelings and emotions?

      (2) Did expression of feeling help establish and develop relationships?

   (3) Were feelings authentic or forced?
   (4) Are participants able to express themselves spontane-
        ously?
c. *Support*
   (1) Was there an adequate climate of respect, acceptance,
        support?
   (2) Were members active in giving support or is the
        climate of support principally a permissive, passive
        thing?
   (3) Did the group prevent any member from clawing at
        anyone?
d. *Confrontation*
   (1) Were people willing to challenge one another?
   (2) Did members confront one another because they cared
        about one another and wanted to get involved?
   (3) Was there any degree of merely punitive confrontation?
   (4) Is conflict allowed in the group? Is it dealt with crea-
        tively or merely allowed to degenerate into hostility?
   (5) Is confrontation really an invitation to another to move
        into the group in a more fruitful way? Do the members
        take the initiative to invite one another into the group
        in various ways?
e. *Response to confrontation*
   (1) Did members reply to responsible confrontation by self-
        exploration rather than defensiveness or counterattack?
   (2) If the person confronted found it difficult to accept
        what he heard, did he check it out with other members
        of the group? Did the other members take the initiative
        to confirm confrontation without "ganging up" on the
        one being confronted?
7. *Trust*
   a. Is the level of trust deepening in the group?
   b. Do members say in the group what they say outside?
   c. If there are problems with trust, do the members deal with
       them openly?
8. *Nonverbal communication.* What do members say nonverbally
   that they don't say verbally (concerning their anxiety, boredom, withdrawal, and so on)?
9. *Leadership*
   a. Does the facilitator model contractual behavior?

b. Was the facilitator acting too much like a leader—that is, trying to get others to do things rather than doing things with others?

c. Is leadership becoming diffused in the group? Or are the members sitting back and leaving most of the initiating to the facilitator? Who are those who are exercising leadership?

d. If necessary, did the facilitator see to it that no one became the object of destructive behavior on the part of others?

10. *Exercises* (if any)

a. If used, were they appropriate? Did they fit into what was happening?

b. Were they well introduced? Were the instructions clear?

c. Were they forced upon an unwilling group?

d. Is there too much dependence on exercises?

e. Does the group always flee exercises even though they might be helpful?

f. Did the exercises used accomplish their goals?

11. *Anxiety*

a. What is the anxiety level of the group? Too high? Too low?

b. Is there always some motivating tension or is the group too comfortable?

12. *Modes of flight and problematic interactions*

a. What are the principal ways in which the group as a whole took flight?

b. In what ways are individuals resisting the process of the group?

c. Do members continue to claim that "they don't know what to do"?

d. *Analysis.* Do members spend a great deal of time analyzing past interactions (an ounce of interaction followed by a pound of analysis)?

e. *Interpretation.* Do members tend to interpret and hypothesize about one another's behavior instead of meeting one another directly?

f. *Quiet members.* Did quieter members move into the group on their own initiative? If not, how was the problem handled? Do individuals or the group rationalize nonparticipation?

g. *Control*. Are there members who control the group by specific behaviors (by always having the focus of attention on themselves, by cynicism, by hostility, by silence, or by some other behavior)?

h. *Pairing*. Were coalitions formed that impeded the progress of the group?

i. *Tacit decisions*. Has the group made any tacit decisions (such as not to discuss certain subjects, not to allow conflict, not to get too close) that affect the quality of the interaction?

j. *Dealing-with-one*

(1) Does the group tend to deal with one person at a time?

(2) If so, is that person usually consulted about being the center of attention for an extended period of time?

(3) Does dealing-with-one mean that others may not contact one another until the person in the focus of attention is "finished"?

(4) Do some people withdraw from the interaction when one person is dealt with for an extended period of time?

(5) If this is a problem for the group, is it dealt with openly?

13.   What is needed to improve the quality of the group?

Obviously other kinds of checklists could be drawn up for different kinds of groups. The checklist should be related to the goals—overriding, procedural, and interactional—of the group.

References

Beier, E. G. *The silent language of psychotherapy.* Chicago: Aldine, 1966.

Bennett, C. C. What price privacy? *American Psychologist,* 1967, 22, 371–376.

Bugental, J. F. T., & Tannenbaum, R. Sensitivity training and being motivation. In E. H. Schein & W. G. Bennis (Eds.), *Personal and organizational change through group methods: The laboratory approach.* New York: Wiley, 1965. Pp. 107–113.

Carkhuff, R. R. *Helping and human relations.* Vols. I & II. New York: Holt, Rinehart and Winston, 1969.

Carkhuff, R. R. *The development of human resources.* New York: Holt, Rinehart and Winston, 1971.

Drucker. P. F. *The age of discontinuity: Guidelines to our changing society.* New York: Harper & Row, 1968.

Egan, G. *Encounter: Group processes for interpersonal growth.* Monterey, Calif.: Brooks/Cole, 1970.

Egan, G. *Encounter groups: Basic readings.* Monterey, Calif.: Brooks/Cole, 1971.

Gibb, J. R. Climate for trust formation. In L. P. Bradford, J. R. Gibb, & K. D. Benne (Eds.), *T-Group theory and laboratory method.* New York: Wiley, 1964. Pp. 279–309.

Herzberg, F. One more time: How do you motivate employees? *Harvard Business Review,* 1968, 46, 53–62.

Howard, J. *Please touch: A guided tour of the human potential movement.* New York: McGraw-Hill, 1970.

Jourard, S. *The transparent self.* New York: Van Nostrand Reinhold, 1964. (Now in a revised version, 1971.)

159

Jourard, S. *Disclosing man to himself.* New York: Van Nostrand Reinhold, 1968.

Jourard, S. *Self-disclosure: An experimental analysis of the transparent self.* New York: Wiley-Interscience, 1971.

Keniston, K. *The uncommitted: Alienated youth in American society.* New York: Harcourt Brace Jovanovich, 1965.

Lynd, H. M. *On shame and the search for identity.* New York: Science Editions, 1958. (Reprinted by Harcourt Brace Jovanovich, 1970.)

Maslow, A. H. *Toward a psychology of being.* (2nd ed.) New York: Van Nostrand Reinhold, 1968.

McLuhan, M. *Understanding media: The extensions of man.* New York: McGraw-Hill, 1964.

Mowrer, O. H. Loss and recovery of community: A guide to the theory and practice of integrity therapy. In G. M. Gazda (Ed.), *Innovations to group psychotherapy.* Springfield, Ill.: Charles C. Thomas, 1968a.

Mowrer, O. H. New evidence concerning the nature of psychopathology. In M. J. Feldman (Ed.) *Studies in psychotherapy and behavioral change.* Buffalo, N.Y.: University of New York at Buffalo, 1968b. Pp. 113–193.

Index